Environmental assessment

Ross Singleton, Pamela Castle and David Short

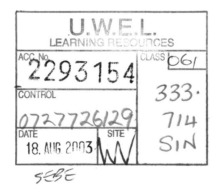

Thomas Telford

Published by Thomas Telford Publishing, Thomas Telford Ltd, 1 Heron Quay, London E14 4JD.
URL: http://www.t-telford.co.uk

Distibutors for Thomas Telford books are
USA: ASCE Press, 1801 Alexander Bell Drive, Reston, VA 20191-4400
Japan: Maruzen Co. Ltd, Book Department, 3–10 Nihonbashi 2-chome, Chuo-ku, Tokyo 103
Australia: DA Books and Journals, 648 Whitehorse Road, Mitcham 3132, Victoria

First published 1999

Cover picture courtesy of David Jarvis Associates

A catalogue record for this book is available from the British Library

ISBN: 0 7277 2612 9

© The Authors and Thomas Telford Ltd

Throughout this book the personal pronouns 'he', 'his', etc. are used when referring to 'the Secretary of State', 'the Minister' etc. for reasons of readability. Clearly it is quite possible these hypothetical characters may be female in 'real-life' situations, so readers should consider these pronouns to be grammatically neutral in gender, rather than masculine, in all cases.

Typeset by The Midlands Book Typesetting Company.
Printed and bound in Great Britain by MPG Books Ltd, Bodmin, Cornwall

Abbreviations and definitions

Act	The Town and Country Planning Act 1990
AGLV	Area of great landscape value
Amended EA Directive	The EA Directive as amended by the 1997 Directive
Annex I projects	Projects listed in Annex I of the EA Directive
Annex II projects	Projects listed in Annex II of the EA Directive
AONB	Area of outstanding natural beauty
APC	The air pollution control regime under Part I of the EPA 1990
BATNEEC	Best available techniques not entailing excessive cost
Blue Book	*Environmental Assessment — A Guide to the Procedures*, Department of the Environment and Welsh Office
BPFO	Best practicable environmental options
Circular	Department of the Environment Circular 15/88 on Environmental Assessment
Contaminated Land Regime	The new contaminated land regime inserted into the EPA by section 57 Environment Act 1995
DG	Directorate-General of the European Commission
DoE	Department of the Environment for England and Wales
DETR	Department of the Environment, Transport and the Regions
Duty of Care	The Duty of Care as respects waste under Section 34 of the EPA

EA	Environmental assessment
EA Directive/EIA Directive	Council Directive of 27 June 1985 on the Assessment of the Effects of Certain Public and Private Projects on the Environment 85/337/EEC
EAP	Environmental action plan
EA Regulations	The Town and Country Planning (Assessment of Environmental Effects) Regulations 1988 (as amended)
ECJ	European Court of Justice
EC Treaty	Treaty of Rome 1957
EIA	Environmental impact assessment
EMP	Environmental management plan
EMS	Environmental management system
EN	English Nature
EPA	Environmental Protection Act 1990
ERP	Environmental response plan
GDPO	Town and Country Planning (General Development Procedure) Order 1995 (SI 1995 No. 419)
GPDO	Town and Country (General Permitted Development) Order 1995 (SI 1995 No. 418)
HGV	Heavy goods vehicle
HSE	Health and Safety Executive
IEA	Institute of Environmental Assessment
IHT	Institute of Highways and Transportation
IPC	The integrated pollution control regime under Part I of the EPA 1990
IPPC Directive	EC Directive on Integrated Pollution Prevention and Control (96/61/EC)
K&D	PB Kennedy and Dokin Ltd
LI	Landscape Institute
LPA	Local planning authority
Maastricht Treaty	The Maastricht Treaty on European Union 1992
MBE	Motherwell Bridge Envirotec Ltd.
NGO	Non-governmental organisation
NRA	National Rivers Authority
PPG 23	Department of the Environment Planning Policy Guidance Note No. 23 – Planning and Pollution Control

RQO	River quality objective
SEA	The Single European Act 1986
Secretary of State	Secretary of State for the Environment
SIA	Social Impact Assessment
SSSI	Site of special scientific interest
SWW	South West Water
TIA	Traffic impact assessment
TEQ	Toxic equivalent
USEPA	United States Environmental Protection Agency
WIA	Water Industry Act 1991
WRA	Water Resources Act 1991
www	World Wide Web
ZVI	Zone of visual influence
1997 Directive	Council Directive 97/11/EC of 3 March 1997 amending Directive 85/337/EEC on the assessment of the effects of certain public and private projects on the environment

Contents

Chapter 6. Case studies   **186**

1

Introduction

The last 20 years have seen an unprecedented increase in land development which is likely to continue for the foreseeable future, and with which has come an increasing concern for the environment — and specifically, the impacts that occur as a result. New developments bring added pressures on existing land uses. Such concerns and pressures resulted in the introduction in 1985 of a European Community Directive requiring the environmental impact of large developments to be assessed as part of the planning process (the 'EA Directive'). While the planning control system since its inception has given some consideration to the environmental impact of development and land use change in a broad sense, the introduction of mandatory requirements to undertake environmental impact assessments for specific types of development has given the process added impetus.

The EA Directive is based on the preventative approach to environmental protection. It requires member states to implement legislation to ensure that, before development consent is given, decision-making authorities are aware of the impact that proposed development will have upon the environment. As such, the EA Directive creates procedural rather than substantive obligations and leaves it open to the decision-making authority whether to actually sanction a project, however damaging to the environment it may be. The EA Directive has been implemented in the UK by legislation which, for certain types of development, requires that environmental information be provided to and considered by the decision-making authority. A pivotal part of this information-gathering process is the preparation of an environmental statement.

1

In the UK alone, hundreds of environmental statements are now prepared each year for developments that either have the potential for creating a significant impact on the environment or for which the examination of the interrelationships between a large number of factors is required. In such circumstances, the normal planning and pollution control regulations would be insufficient to address such issues, but techniques of environmental assessment can be used to address and assess the interdisciplinary facets of a development proposal.

The rise in the number of environmental statements being prepared has resulted from the increase in the number of developments for which the submission of such a statement is a legal minimum, but it is also a reflection in public concern. The preparation and submission of voluntary environmental statements, for example, represents the desire of many companies to inform the public and to be 'seen to be green'. Such practices are becoming increasingly popular as companies realise the public relations value of keeping the public informed and creating a 'green' image.

Techniques utilised as part of the environmental assessment process have matured and been refined since the adoption of the EA Directive and its implementation into UK law and, as a consequence, the standard of environmental statements is constantly improving. This is occurring through an increase in the experience and training of the practitioners (numerous training courses are now available and some universities even offer Bachelor of Science degree courses in environmental assessment), through the greater availability, number and usefulness of guidelines recommending methodologies for undertaking assessment of specific issues (such as landscape assessment and the undertaking of ecological surveys that have been produced by both the Government and experienced bodies such as the Institute of Environmental Assessment) and through the greater experience and understanding of the regulatory authorities.

Quality in environmental assessment has attracted close scrutiny and attention in the last five years as techniques and methodologies are refined. Rigorous analysis, responsive consultation and responsible administration (the 'three Rs') have been identified as a cornerstone in achieving quality (Sadler, 1996). Focus on the quality of environmental assessments has occurred of late, partly because of the increasing pressure on practitioners due to competition, budgetary restrictions and the recent worrying introductions of success fees.

A greater understanding by those involved of the issues to be addressed as part of the environmental assessment process has however been achieved, while guidance on how those issues can be impacted upon by certain types of development, and how the statutory consultees prefer to see the information presented, is now available.

The undertaking of an environmental assessment requires the completion of a number of key steps conducted more or less in the same sequence regardless of the type of development being considered. The first stage is generally viewed as the description of the proposed development; this is followed by the examination of the existing environment prior to the third stage, i.e. the forecast of the probable impacts that may occur and the specification/design of measures to mitigate the impact. Included within each of these steps are a myriad of other related activities that are necessary to ensure satisfactory completion of the statement.

Scoping and consultation (from both the public and relevant local authorities) comprise an area that has benefited enormously as a result of the increase in experience and improved practices from environmental assessment exponents; these techniques and approaches have been widely shared in the scientific press and at conferences, resulting in their dissemination to a wide audience. Scoping and consultation have been two areas of environmental assessment that have received little attention in the past, with most statements focusing upon the actual impact assessment/prediction exercises; experience has shown that these two elements are crucially important to the environmental assessment process and as a result the techniques used have become well developed and matured over a relatively short period of time, i.e. the last two to three years.

The fundamental principles of environmental assessment that enable the production of the environmental statement have remained unchanged, however: scoping an assessment is a critical element of the study that requires focused attention and direction that will ultimately lead to the success or failure of the statement; consultation with regulatory bodies and other interested parties is gaining in significance; the limitations of the study must be clarified so that the study boundaries are clearly defined, approaches are justified and all assumptions are clearly stated; assessment of impacts can be undertaken in a qualitative or quantitative manner (depending upon the significance of the issue); prediction of impacts is then possible and significance can be assessed; reporting of the environmental assessment must be undertaken in a manner that is easily understood by the expert and non-expert alike and it should enable reproducibility of the results to be achieved.

There is no best methodology for conducting environmental impact assessments; the variations in the type and characteristics of the development and, perhaps more significantly, the differences between each plot of land being considered for development will result in varying techniques being used throughout the assessment process. This book contains many examples of best practice for a variety of circumstances, and more specifically provides examples of useful techniques that can be used in the gathering, analysis and communication of information about the consequences of development activities. However, the overall methodology that should be applied in almost every case, and one that is recommended here, is to consider each specific management issue and to address, through design, the environmental assessment process to fit the need and the time and resources that are available.

The aim of this book, therefore, is to impart some of this experience to individuals involved in undertaking environmental assessments so that they can benefit from its content and thus contribute to the improvement in environmental assessment practices and techniques. It is intended to provide a source of practical information and guidance on why environmental assessments are necessary, what they are intended to achieve, what the legal requirements are and how the assessments should be carried out. Environmental assessment is a relatively new area of work; very valuable and specific guides are available for particular types of development and for particular subject areas but there are only a few instances of practical guidance on how the environmental assessment process should be approached and completed.

This book is not meant to be a fully inclusive manual on how to undertake an environmental assessment for all types of development in all environmental areas; such a document would be extremely lengthy (if in fact it could be produced at all) and an exercise in futility, given that the basis behind environmental assessment is that it is a site-specific exercise and one that must be undertaken in a manner that reflects the development type and site environment parameters.

The specific audience at which this book is aimed are environmental/ pollution science undergraduates, civil engineers involved with development projects, assistant/junior environmental consultants and those charged with the responsibility of reviewing completed documents, i.e. those undertaking or involved in the environmental assessment process. It is likely that some practitioners of environmental assessment will not find sufficient detail within this book for conducting a specific type of data collection exercise or modelling methodology, but it will provide them with a first point of reference from which to cross-refer to other specific subject areas.

The structure of this book is as follows: Chapters 2, 3 and 4 relate to the legal aspects of environmental assessment and place them in the context of environmental engineering and development. They first provide a broad perspective on environmental legislation, before examining the legislative structure and requirements of the environmental assessment process and its role within the planning control system.

Chapter 2 is divided into two parts. The first gives an introduction to European Community (EC) environmental law and policy, the breeding ground from which the EA Directive was spawned. The second gives a broad overview of environmental law in the UK, much of which has been enacted to implement the requirements of EC legislation.

Chapter 3 discusses the provisions of the EA Directive adopted by the European Community in 1985 and also comments on the amendments that were made in 1997, which have yet to be implemented in UK law. This chapter also examines the progress that has been made to date in implementing the requirements of the EA Directive in UK law and suggests how the 1997 amendments will affect the environmental assessment process in the UK in the future.

Chapter 4 focuses on the legislative provisions which govern the environmental assessment process in the UK at present. It sets out in some detail the requirements of the Town and Country Planning (Assessment of Environmental Effects) Regulations 1988, which apply to the majority of projects requiring environmental assessment in the UK. Consideration is also given to the requirements for environmental assessment in relation to projects which do not fall under the town and country planning system and which require development consent from some decision-making body other than a local authority.

Chapter 5 presents a review of the methodologies used in environmental assessment and provides guidance on the techniques to be used, the pitfalls to be avoided and the methods that can be adopted that will significantly aid the preparation process. Where relevant, examples of real situations are presented in order to demonstrate or illustrate a point.

The presentation of case studies that examine the issues addressed in a number of environmental statements can be found in Chapter 6. These are summaries of actual environmental statements that have been submitted as part of the planning application process.

The full text of parts of the legislative provisions are included in the Appendices, so that the book can be used as a single point of reference or as an initial source from which further information and guidance can be identified. In particular, Appendix 5 presents the results of research examining the use of the Internet and the World Wide Web in environmental assessment and contains a list of many useful web pages from which data can be obtained.

Environmental assessment is a rapidly changing process and one that will mature more fully over the next ten years. Changes to the EA Directive were made in 1997 which require member states to amend their implementing legislation by 14 March 1999 in a number of ways. This will lead to certain significant changes to the environmental assessment process in the UK, including a broader range of projects that will be subject to the mandatory requirements for environmental assessment, and further provisions to encourage scoping studies. Similarly, several guidelines on the undertaking of specific assessment studies are expected to be published in the near future which will update the accepted methodologies that should be used. It is expected that there will be a greater reliance upon new techniques over the next ten years; the development of more sophisticated and accurate modelling tools, availability of reliable and accurate environmental data and the use of geographical information systems will lead to changes in the way that assessments are approached, undertaken and completed.

Similarly, the development of audit monitoring as a specific requirement is likely to become the norm and a host of other 'improvements' will be developed that will change the appearance of environmental assessment as it is now known.

The Authors

The book has been co-authored by practitioners involved in both the legal side and practical side of environmental assessment. Chapters 2, 3 and 4 were written by Pamela Castle and David Short of Cameron McKenna. Chapters 1, 5, 6 and Appendix 5 were written by Ross Singleton of PB Kennedy and Donkin Ltd.

Ross Singleton is head of the environmental assessment team of PB Kennedy and Donkin Ltd (formerly Rust Environmental), a company which has undertaken over 60 environmental assessments for a wide variety of development types. Following degrees in environmental science and public health engineering, Ross specialised in environmental assessment in Hong Kong, working on many of the Territory's major infrastructure development projects in the late 1980s and early 1990s. Since returning to the UK and to PB Kennedy and Donkin Ltd he has worked extensively in the UK, Europe and the Far East on environmental assessment. His contribution to this book comes as a result of witnessing first-hand the development and maturing of environmental assessment over the last ten years.

Pamela Castle is Partner and Head of the Cameron McKenna Environmental Law Group, a specialised group which monitors the development of environmental legislation on a global basis, but with a particular emphasis on the UK and the European Union. She holds an honours degree in chemistry and worked in industry for a number of years before qualifying as a solicitor. She has been responsible for advising numerous clients with 'household names' on all aspects of environmental protection legislation in both contentious and non-contentious matters. She is well known as a public speaker and sits on a number of committees, such as those of the Environment Agency, the National Radiological Protection Board and the World Wide Fund for Nature (UK).

David Short is a member of the Cameron McKenna Environmental Law Group. After gaining a Bachelor's degree in law at the University of Oxford, he completed a Master's degree in international and comparative law at McGill University in Montreal, focusing on international environmental law. He has also undertaken post-graduate research work into international environmental dispute resolution at the McGill Law Faculty. David now specialises in environmental law at Cameron McKenna's London office.

Acknowledgements

Ross Singleton
As the undertaking and preparation of an environmental assessment is a team activity involving specialists in many areas, so the author has used the experience and knowledge of a large number of specialists. It is therefore appropriate and necessary to acknowledge the contribution of my colleagues both within PB Kennedy and Donkin and at a variety of other specialist companies. My thanks go to the following for their help and assistance in the preparation of this book (PB Kennedy and Donkin unless otherwise stated): Andy Bartlett (noise), Alec Jeffries (highways), Mabel Munoz-Devesa (WWW research and proof reading), Johnathon Larkin (ground water modeling), Bill Finlinson (water resources), Peter McKendry (waste management), Michaela Bergman of Gilmore Hankey Kirke (Cultural and social) and Jim Meadowcroft of David Jarvis Associates (landscape and visual). My thanks also go to the two companies (Motherwell Bridge Envirotec and South West Water) who allowed us to include details of their projects as case studies in this book as well as to the various companies/individuals who allowed reproduction of photographs etc.

I would also like to acknowledge the considerable guidance in environmental assessment provided by Bob and Topsy (though I didn't realise it at the time) and to express my thanks to them and also to my wife, Cath, who has tolerated my long and regular absences during the lengthy preparation of this book.

Pamela Castle and David Short

We would like to acknowledge the support of the other members of the Cameron McKenna Environmental Law Group and to give special thanks to Ann Peirson-Hills and Mark Challis for their substantial help and assistance in the preparation of this book.

We would like to thank Her Majesty's Stationery Office for permission to reproduce parts of The Town and Planning (Assessment of Environmental Effects) Regulations 1988 which appear in the Appendices.

2

Environmental law — the broader perspective

THE EUROPEAN PERSPECTIVE

European Community environmental policy

The European Community is the main source of environmental law in the UK. In particular, it is the source of the law and policy on environmental assessment which is the subject matter of this book. It is necessary at the outset to appreciate, however, that environmental law in the European Community is not just about legal rules. Much environmental law should be seen in the context of policies and principles established by the European Community in applying its laws and carrying out government. These policies and principles may be generally applicable for all issues across the European Community or specifically applicable to environmental issues.

The Treaty of Rome and the Community Action Programmes

The Treaty of Rome 1957 (hereafter referred to as the 'EC Treaty'), which created the European Economic Community, now renamed the European Community, contained no reference at all to the environment. It was not until 1972, at the Paris Summit, that the need for a European Community policy on the environment was acknowledged. This led, in 1973, to the first of a series of Community Action Programmes on the Environment, which formed the first European Community environmental policy. There have now been a total of five Community Action Programmes on the Environment, each of which covers a period of approximately five years. The current Action Programme, adopted in 1992, is entitled 'Towards Sustainability — a European Community Programme of Policy and Action in Relation to the Environment and Sustainable Development' and constitutes the environmental policy of the European Community until the year 2000. The Community Action Programmes provide a framework for environmental policy and set out the areas on which the Commission intends to legislate. They do not themselves have any binding legal force.

The framework set out in each Community Action Programme on the Environment has two main objectives: first, to identify specific legislative proposals which the Commission intends to initiate and develop; second, to introduce and discuss ideas and issues relating to environmental policy, including possible future developments. The general principles of environmental policy contained in the Action Programmes which are of relevance to environmental assessment can be summarised as follows:

- that prevention is better than cure;
- that environmental effects should be taken into account at the earliest possible stage in decision making;
- that the exploitation of natural resources which causes significant damage to ecological balance must be avoided;
- that scientific knowledge and understanding must be developed in order to enable effective action to be taken;
- that activities carried out in one member state must not cause deterioration of the environment in another.

In addition to the general policy set out in the Community Action Programmes described above, the European Community also makes more specific environmental policy, for example in respect of waste and energy matters.

The EC Treaty has now been amended by the Single European Act 1986 ('SEA') and the Treaty on European Union 1992 ('Maastricht Treaty') to include specific environmental provisions. Article 2 of the EC Treaty is now amended to include, as one of the fundamental objectives of the Community, the promotion of 'sustainable and non-inflationary growth respecting the environment' and Article 3 is amended to include as one of the activities 'a policy in the sphere of the environment'. Furthermore, Articles 130R, 130S and 130T have been inserted to provide a clear legal basis for all EC environmental policy, whether contained in Community Action Programmes on the Environment or in policy upon specific environmental issues.

Article 130 of the EC Treaty

The objectives of any European Community action relating to the environment are set out in Article 130R(1) of the EC Treaty (as amended). There, four key objectives appear:

- to preserve, protect and improve the quality of the environment;
- to contribute towards protecting human health;
- to ensure a prudent and rational utilisation of natural resources;
- to promote measures at an international level to deal with regional or worldwide environmental problems.

With the key objectives in mind, any action by the European Community relating to the environment must take into account the core principles which are set out in Article 130R(2). They are:

- that preventative action should be taken (in preference to remedial measures);
- that environmental damage should as a priority be rectified at its source; and
- that the polluter should pay for pollution and its effects.

In addition, the EC Treaty now specifically states that 'environmental protection requirements must be integrated into other Community policies' and not just environmental policy (Article 130R(2)).

European Community legislation

An introduction to EC law in the UK

In 1973, the UK became a member of the European Communities, of which there were in fact three, namely the European Economic Community, the European Coal and Steel Community and the European Atomic Energy Community. (The Maastricht Treaty changed the official name of the European Economic Community to the European Community ('EC'), thus changing the EEC Treaty into the EC Treaty. It should also be noted that the Maastricht Treaty introduced the term European Union ('EU') to describe the umbrella identity which consists of the three European Communities plus two extra areas of intergovernmental co-operation. This term is not strictly applicable to matters of law relating to the EC Treaty although it may be properly used in certain cases, such as to describe 'EU citizens'. For the sake of simplicity and clarity, the term EC has been used throughout in this book).

To give effect to its legal and political commitment in joining the

11

European Communities, the European Communities Act 1972 was enacted by Parliament. It provides that, without the need for any further legislation, all the provisions of the EC Treaties governing the European Communities and the rules made under them are to be given the force of law in the UK, to be recognised and available in law, and to be enforced and followed (European Communities Act 1972 Section 2(1)).

The fundamental principle is that EC law has primacy over conflicting national laws and supersedes all national laws that diverge from it. This means in general terms, that:

- EC legislation may in certain circumstances be directly effective, whether or not there has been implementing legislation in the UK; and
- if there is a conflict between EC and UK legislation — even a UK Act of Parliament — the EC legislation takes precedence.

A number of types of legislative measures can be made by the EC institutions under the EC Treaty. It is important to understand what each type of measure sets out to achieve and the legal force that each has; these aspects are discussed in the following paragraphs.

Directives

Directives are the principal type of EC legislation used for environmental issues and the protection of the environment. They come in two varieties. They may be general in nature, stating policies and objectives or aims that the Council wishes to achieve; or they may be more specific, requiring the member states to meet specific targets or standards. Directives of the first type are often called Framework Directives; and Directives of the second type, Daughter Directives, address or set out more precisely the specific objectives falling within the relevant Framework Directive.

Directives require member states to achieve certain prescribed results within a stated period of time, but leave to the member state the methods for implementing the measures that will achieve that result. Failure by a member state to implement a Directive into its own national law may result in the European Court of Justice (ECJ) taking proceedings against it.

Provided that the definitions in, and the requirements of, a Directive are sufficiently precise, it can have direct legal effect as soon as the time has expired for the member state to implement it. The importance of Directives with direct effect is discussed in Chapter 3. Generally, however, the doctrine of direct effect means that the provisions of a Directive which are not transposed into national laws may be directly enforceable against the government of a member state. Further, individuals and companies may be able to enforce a Directive against government bodies and public authorities where there has been a failure to enact the necessary implementing rules within the prescribed period of time.

In the UK, implementation of Directives is usually by means of Acts of Parliament (primary legislation) or Statutory Instruments (secondary legislation). Administrative circulars have also been used to implement binding requirements of Directives. It is beyond the scope of this book to consider the detail of the large number of Directives on environmental protection, but in Table 1 they are summarised and divided into various categories.

Regulations

Regulations made by the EC are legislative acts of general application. They are distinguishable from Directives in that they are normally directly applicable in and binding upon all member states and their citizens without the need for any national implementing legislation. Regulations are used when the EC considers that the matter to be regulated should offer no discretion to member states as to the method of implementation, in the interests of complete harmonisation throughout the Community. Additionally, they have the advantage of avoiding the delays that almost always occur where national implementing legislation is needed to implement a Directive. For these reasons, Regulations have been increasingly employed by the Commission for legislating on environmental matters such as:

- the phasing out of chloroflurocarbons (CFCs) which deplete the ozone layer;
- the supervision and control of waste shipments within, into and out of the EC;
- the evaluation and control of environmental risks from chemical substances; and
- voluntary ecolabelling and environmental management schemes.

Decisions

Decisions by an EC body are legally binding on those to whom they are addressed, whether this is to member states, companies or individuals. They play only a small part in environmental matters at the EC level.

13

Table 1. Summary of issues addressed by European Community Directives on environmental protection

Category and aim	Examples of scope
Water	
Setting quality standards for different water uses	Drinking water; bathing water; fishery waters
Setting emission standards/limits for certain discharges to water	Hazardous substances into groundwater; mercury; cadmium
Air	
Setting quality standards for air	Sulphur dioxide; nitrogen oxides; lead
Setting standards for emissions to air	From municipal waste incinerators, large combustion plants, industrial plants, vehicles
Waste management	
Control of waste activities — waste minimisation, recycling, recovery, disposal	General waste; hazardous waste; packaging and packaging waste
Dangerous substances	
Storage and use of hazardous materials and the control of hazardous activities	General and industry-specific
Setting of product standards	Lead in petrol; classification, packaging and labelling of dangerous substances
Nature protection	
Protecting wild animals and natural habitats	Wild birds; whales; endangered species; the Antarctic
Noise	
Setting emission standards for noise from vehicles and machinery	Motor vehicles; aeroplanes; construction plant
Nuclear safety and radiation protection	
Setting safety standards and providing for public access to information	General public; workers; patients; food

The European Community institutions

In order to understand the working of EC legislation and its effect in the UK, the functions of the EC institutions need to be explained briefly.

The Council

The Council, which since the Maastricht Treaty has been officially called 'The Council of the European Union', is a political body consisting of one representative from each member state and is the principal decision-making body of the EC. It has a primary role in law making and no legislation can be adopted against its will. It is not a fixed body, but has a fluctuating membership depending on the subject under discussion. Each member government has one seat on the Council and will usually send the Minister with responsibility for the matter to be agreed, for example the Chancellor for matters relating to economic policy or the Environment Minister for environmental matters. At least twice a year there is a meeting of a specially constituted body, termed the 'European Council', which consists of the heads of state or of government of the member states and the President of the Commission, and which acts on matters of both EC law and political co-operation.

It is common, particularly when non-contentious issues are being discussed by the Council, for governments to send senior civil servants to negotiate on their behalf rather than sending the elected members of government. Meetings of the Council are also attended by the Commissioner responsible for the subject under discussion. The Commissioner is entitled to participate in the discussion and is usually influential in persuading members to agree on the legislative proposals being aired before them. Meetings are chaired by the member state which holds the presidency of the Council. The presidency is rotated every six months and the presiding state sets the agenda as well as chairing meetings.

The proposals debated within the Council are put forward by the Commission. The Council may decide to adopt a proposal unamended, to amend and then adopt it, or to reject it altogether. Often it will debate a proposal but defer a decision to allow further discussion, either at EC level or within the member states. Depending on the subject matter and the Article of the EC Treaty on which the legislative proposals are based, decisions of the Council are taken either by a qualified (i.e. weighted) majority vote or by a unanimous vote. The voting procedure adopted is often crucial. In relation to environmental measures, it has not always been clear which Article of the EC Treaty (as amended) constitutes the correct legal basis for legal instruments, and therefore which voting procedure is applicable. This has led to disputes between the Commission, the Parliament and the Council. For example, Directive 89/428/EEC (on waste from the titanium dioxide industry) had to be replaced after it was annulled by the European Court of Justice in 1991. Annulment was ordered on the ground that this Directive was adopted on the wrong legal basis, namely Article 130S (protection of the environment, which required a unanimous vote of the Council) instead of Article 100A (completion of the internal market, which required only a qualified majority vote of the Council, but allowed greater input from the Parliament through the co-operation procedure). In the environmental arena contentious draft Directives often attract much debate and discussion between the EC institutions and member states. It can often take as long as five years or more from the date of the official proposal for a Directive or Regulation until its formal adoption.

The Maastricht Treaty seeks to resolve some of this conflict, although it does make the legislative process even more complex. For the adoption of environmental legislation, qualified majority voting is now the standard procedure in the Council. However, a unanimous vote is still required in several areas, including measures concerning town and county planning, fiscal measures (such as the proposed carbon tax), water resource management and energy supply.

The qualified majority vote operates by allotting more votes to the larger members states (up to ten) than to the smaller member states (Luxembourg has the least, with two votes). For a measure to be adopted by a qualified majority, 62 votes out of the total 87 votes are required. This numerical threshold has the effect that two of the largest states or eight of the smallest states can be outvoted on a qualified majority vote.

The Commission

The Commission is essentially the civil service and central bureaucracy of the European Community, based in Brussels. It is headed by 20 Commissioners drawn from all 15 member states. The five member states with the largest populations have two Commissioners each. Unlike members of the Council, Commissioners do not represent the national interests of the member states. They must act independently and work for the general good of the EC. The Commission initiates policy and legislation in the form of proposals which are then laid before the Council (see above) for debate and eventual adoption. The Council, acting either alone or with Parliament, is the only body which can adopt legislation, although in relation to technical and administrative matters it may — and does — delegate legislative powers to the Commission.

The Commission executes decisions made by the Council and oversees the daily running of EC policies and legislation. It also has an express duty to ensure that EC law is enforced (implemented and complied with) in the member states and may take proceedings against them if it is not. It is the Commission which proposes and drafts new environmental legislation and the Community Action Programmes on the Environment.

The Commission is divided into 24 departments called Directorates-General (usually referred to as 'DGs' for short). Each DG has delegated responsibility for policy, legislation and enforcement of a specific topic or area. Each DG is headed by a Commissioner and one Commissioner may be responsible for more than one DG.

The environment (as well as nuclear safety and civil protection) is dealt with by DGXI which was set up in 1981. It is the responsibility of DGXI to initiate and follow through proposals for new environmental legislation. Officials within DGXI draft legislative proposals for environmental protection, following representations made by political and business interests such as trade associations at all levels. DGXI is composed of five Directorates. Directorate A deals with general and international affairs, Directorate B with environmental instruments, Directorate C with environmental safety and civil protection, Directorate D with environmental quality and natural resources, and Directorate E with industry and environment. Directorate B has responsibility for the issue of environmental assessment. As other policy areas have some impact on the environment, other DGs may have some responsibility for certain environmental issues.

The Parliament

The European Parliament is the central elected Parliament of the European Community. Since EC law is for the most part made by the Council, the Parliament does not have legislative powers in the same way as a national parliament. It is mainly a consultative and advisory body, although it does have certain supervisory powers over the Commission and the budget. In addition, its power to influence legislation has been enhanced over the years through the introduction of the 'co-operation procedure' by the SEA and the 'co-decision procedure' by the Maastricht Treaty. There are now few texts of any significance which can be adopted without at least the Parliament's opinion first having been sought. Members of the European Parliament (MEPs) are directly elected by the citizens of each of the 15 member states from which they come, for a term of five years. Each member state is allocated a fixed number of seats, larger countries having more seats than smaller ones. The European Parliament is one of the largest assemblies in the world, with over 600 members.

While the Commission and the Council are the main legislators in the European Community, the European Parliament's role as a legislative body has been extremely limited. Before the Maastricht Treaty it had limited capacity to influence the legislative process during the 'consultation' procedure, under which the Parliament was required to give its opinion on proposed legislation before the Council was able to adopt it. There were several occasions on which Parliament's views were followed, despite there being no duty to do so.

The SEA and the Maastricht Treaty enhanced the Parliament's powers, giving it greater opportunity to amend draft legislation under the 'co-operation procedure' and a new right to block measures under the 'co-decision procedure' in certain defined areas. The powers of the Parliament now extend to the ability to veto legislation approved by the Council under this latter procedure.

The Parliament also has a number of permanent committees which examine and amend draft legislation. Environmental legislation is dealt with by (*inter alia*) the Committee on the Environment, Public Health and Consumer Protection.

The European Court of Justice

The European Court of Justice ('ECJ') has no part to play in the legislative process — it is the judicial body of the European Community. All ECJ judgements are binding upon the European institutions, member states and individuals, and decisions exert a strong influence on EC policy. The ECJ comprises 15 judges appointed by agreement of the member states. It is the supreme authority on matters of EC law, including the interpretation of the EC Treaty and EC legislation and implementation of EC law by member states. Its functions are:

- to give preliminary rulings on questions of law referred to it by national courts;
- to determine, at the instance of the Commission or of any member state, the validity of acts of the European Institutions or of any other member state;
- to hear appeals from decisions of the Commission.

There is a Court of First Instance attached to the ECJ with jurisdiction to hear appeals from decisions of the Commission on competition law, 'staff' cases and certain matters arising under the European Coal and Steel Treaty.

Previously the ECJ's powers in relation to complaints of failure by member states to transpose directives properly into national law were merely declaratory. Therefore the sanction of the ECJ often had no more effect than to influence domestic courts in matters of enforcement of EC environmental law. However, the Maastricht Treaty amended Article 171 of the EC Treaty so as to empower the ECJ to impose a lump-sum penalty on a member state which fails to comply with its judgement.

The Economic and Social Committee

This institution plays a consultative role in EC decision making and its members are appointed by the Council. They represent a variety of interests such as trade unions, employers' federations, industry, agriculture, consumer bodies, professional organisations and the general public. Environmental interests are represented. The Committee is entitled to advise the Community institutions of its own initiative on all questions affecting EC law. In addition the Commission and Council may consult the Committee whenever they consider it appropriate, and in some cases they are obliged to do so as an essential procedural requirement under the EC Treaty. However, as the Committee's powers are purely advisory, any hostile opinion it may give can be ignored by the other Institutions if they see fit to do so.

The European Community legislative process

The procedure by which European legislation, such as Directives and Regulations, is made is complex. Legislation is initiated by a proposal for a Directive or a Regulation put forward by the Commission. However, there are provisions enabling the Council to request the Commission to undertake studies and submit appropriate legislative proposals.

Before the Commission issues a proposal for legislation, a process of consultation and negotiation takes place on both a formal and an informal level, involving organisations representing various interests such as industry, consumers, environmentalists and national experts. Formal consultation occurs through the EC-registered umbrella organisations, such as the European Environmental Bureau, an organisation which represents the environmental interests of non-governmental organisations from 23 countries. Informal discussion takes place with interested parties to develop a view on whether certain areas require legislation and how this legislation may be framed. The Commission may issue consultation documents (usually called 'Green Papers'), setting out its ideas and highlighting problem aspects, perhaps with a proposed draft for a Directive. Additional discussion takes place between governments, industry, consumers and environmental organisations. There is a growing concern, however, that this present system of consultation is inadequate and that the EC bureaucracy is too small to handle the number of matters for which it has responsibility.

After the consultation process, the Commission puts forward a formal proposal for legislation. This will generally be published in the 'C' series of the *Official Journal of the European Communities*. Once adopted (by whatever procedure) it is contained in the 'L' series.

The main legal bases for proposed environmental legislation (meaning the Articles of the EC Treaty under which such proposed legislation may be put forward for adoption) are at present Article 100A, which enables Directives to be made that seek to harmonise laws and administrative practices of member states which directly affect the establishment or functioning of the common market, and Article 130S, which provides a specific basis for environmental protection laws. The one most commonly used for environmental legislation is Article 130S. Whether proposed under Article 130S or 100A, most environmental legislation is now subject to qualified majority voting in the Council. The only exceptions are those proposed under Article 130S, which are measures primarily of a fiscal nature, measures relating to town and country planning or land use (unless concerned with waste management), water resource management measures and measures affecting national policies on energy supply. Adoption of these measures still requires a unanimous decision of the Council; the Parliament has only a weak influence over them as they are subject to the old 'consultation procedure', which means that there is no more than a duty to consult the Parliament. Depending on the legal basis, all other proposed environmental legislation will follow one of the two more complicated procedural pathways:

- The co-operation procedure (for legislation proposed under Article 130S). Basically, the Council adopts a 'common position' based on the Commission's proposal, which is then passed on to the Parliament for consideration. If Parliament decides to reject the common position then the Council can only adopt it by a unanimous decision. If Parliament proposes amendments then the proposal goes back to the Commission for consideration. If the Commission chooses to insert the Parliament's proposed amendments then these can be adopted by the Council by a qualified majority. Alternatively, the Council can reject these and adopt its own amendments by a unanimous decision or it can allow the proposal to lapse.
- The co-decision procedure (for legislation under Article 100A). This, the most complex procedure, gives Parliament an absolute

right of veto. The first part of the co-decision procedure is the same as for the co-operation procedure. If, after that, the Council does not succeed in adopting the amended proposal, the matter must be referred to a Conciliation Committee with a view to reconciling the positions of the Council and the Parliament. If a common position cannot be agreed then the Parliament may reject the text completely by an absolute majority of its members.

THE UK PERSPECTIVE

Environmental law in the UK

The UK has a long history of environmental controls. The earliest significant measures were taken under the public health provisions in the 1800s, to deal with the need to protect the health of the people living in increasingly polluted industrial cities. Legislation followed in due course upon air emissions, water emissions, town and country planning and — in the 1970s — waste. As a matter of general approach, however, environmental law sought to address each of the environmental media, namely air, water and land, in a way which was discrete to each medium. It is only comparatively recently that there has been an attempt to integrate the control of pollution across the three media.

Over the last few years, as public concern about the environment has grown, so the focus of environmental law has shifted. Recent legislation, particularly the Environmental Protection Act 1990 ('EPA'), introduces integrated pollution controls. This process has been continued by the unification of regulatory and enforcement controls, previously contained within several separate regulatory authorities, into the all-encompassing Environment Agency. The Agency is a 'one-stop shop' for nearly all environmental policy, legislation and enforcement.

A central feature of modern environmental law is a recognition that reliance upon the enforcement of private rights, such as those under the law of nuisance (discussed below), is insufficient for protecting the environment or for compensating those suffering from the effects of environmental pollution or degradation. This is because much environmental damage does not impinge directly on persons or personal property to which private rights apply. For example, while potentially harming the environment, air pollution or discharges into rivers and streams do not usually harm personal property. Even where personal property or personal damage does occur, such as where trees on private land are damaged by dust or where people suffer from respiratory problems due to vehicle emissions, it is often the case that no single person can be said to be responsible for causing that damage. Indeed, it may have been caused by numerous separate pollutants, from various sources, perhaps over many years. Nowadays sanctions, usually fines, are imposed in the criminal courts for failure to comply with the environmental laws set out by statute. More recently, however, legislative policy has tended towards an increased emphasis on remedying the effects of pollution, in particular the clean-up of contaminated land, by administrative action.

The sources of UK environmental law and liability — an overview

This section describes in general terms the sources of environmental law and liability in England and Wales. In the separate jurisdictions of Scotland and Northern Ireland there are important differences in the form and operation of laws, although only a few in substantial issues of principle. For convenience, the terms 'UK' and 'UK law' are used throughout this book. However, it should be noted that in the context of this book these terms relate only to England and Wales. If specific information on the law in Scotland or Northern Ireland is required, then the specific provisions relating to those jurisdictions should be consulted.

There are two main sources of liability in relation to environmental protection, namely criminal law and civil law. In addition, some of the more recent statutory provisions for the protection of the environment, such as the contaminated land regime, impose what may be called 'administrative liability' (for it may not properly be classified as criminal or civil liability) for the cost of clean-up and remediation of pollution.

Crimes and civil wrongs

In most instances liability under environmental laws will constitute a crime, a civil wrong or both. Any one act or omission may be both a crime and a civil wrong. The distinction between the two is not in the nature of the wrongful act, but rather in the legal consequences and sanctions following from the act. If an act or omission is capable of being followed by criminal proceedings, it is regarded as a crime (i.e. a criminal offence). If the act or omission is capable of being followed by civil proceedings, then it is regarded as a civil wrong. Entirely separate procedures apply in each case, as Table 2 illustrates.

Table 2. Crimes and civil wrongs

	Criminal	Civil
Proponent of proceedings	Prosecutor (Director of Public Prosecutions or regulatory authority/body) prosecutes the defendant.	Plaintiff (an individual sues the defendant.
Result	If proceedings successful, 'conviction' of the defendant who is guilty of an offence.	If proceedings successful, 'judgement' for plaintiff; defendant has committed a civil wrong.
Sanctions	Punished by fine and/or imprisonment as set out for the crime in question in relevant statute; defendant will have criminal record.	Remedy — payment of money (damages); transfer property; defendant prohibited from acting or made to act (injunction) or perform a contract.

Criminal Liability

Introduction

Most contemporary environmental laws involve the regulation of activity by private individuals and bodies, but may extend to the acts of public bodies as well. The rules and standards are set by statute and a failure to comply with them can result in criminal sanctions. Depending upon the offence, the enforcement of a breach of the criminal law involves prosecution by either the appropriate regulatory body, usually the Environment Agency, or the Director of Public Prosecutions through the Crown Prosecution Service. In environmental matters the use of criminal sanctions is often described as the 'regulatory regime' or the 'command and control' approach to environmental protection and pollution control. This section gives an outline of the bodies responsible for the administration and enforcement of environmental legislation followed by a brief description of some of the main areas of the regulatory regime applicable to environmental issues.

Administration and enforcement

The Department of the Environment, Transport and the Regions ('DETR') is the authority responsible for environmental policy and its implementation. Before 1997 there were separate departments and it was the Department of the Environment ('DoE') which was the relevant authority. Regulation of waste-water discharges and waste management is the responsibility of the Environment Agency. The private-sector water and sewerage companies are self-regulatory with respect to discharges to public sewers except in the case of particularly hazardous discharges, which are also regulated by the Environment Agency. Regulation of air pollution is split between local authorities and the Environment Agency. Contaminated land will be regulated by local authorities unless the land is a special site in which case the Environment Agency will assume responsibility. Local authorities have responsibility for noise controls. The Department of Trade and Industry, the Health and Safety Executive and the Ministry of Agriculture, Fisheries and Food also have responsibilities in the environmental area.

The Environment Agency was created on 8 August 1995 and commenced operations on 1 April 1996. It took over the functions of the National Rivers Authority, Her Majesty's Inspectorate of Pollution and the Waste Regulation Authorities, to establish a single body to assume overall responsibility for most areas of environmental protection.

Integrated pollution control and air pollution control

Part I of the Environmental Protection Act 1990 (EPA) introduced a system of integrated pollution control (IPC) and air pollution control (APC). The concept underlying IPC is that emissions to air, water and land should all be regulated by a single permit under the aegis of a single enforcement agency, now the Environment Agency. APC, on the other hand, is a matter regulated by local authorities.

The Environmental Protection (Prescribed Processes and Substances) Regulations 1991 (as amended) (Ref: SI 1991/472 as amended) set out a list of processes, known as prescribed processes, which fall under either the IPC or the APC regime. The scheduled processes are divided into Part A (IPC) and Part B (APC) processes. Part B processes, which are regarded as less seriously polluting ones, are controlled by local authorities. Whichever regime applies, however, an authorisation is required before a prescribed process can be operated. It may be refused, or granted subject to conditions. It is a general condition of all authorisations that the 'best available techniques not entailing excessive costs' must be used to prevent or minimise releases of prescribed substances to air, water and land.

It is an offence for a person to operate a prescribed process without authorisation or to operate it in breach of any conditions attached to that authorisation; the criminal sanctions that apply to such an offence are summarised in Table 3. Further, in addition to or instead of imposing any punishment, the court may order an offender to take specified steps to remedy the matters complained of within a specified time limit.

The IPC system will be substantially amended and extended when the EC Directive on Integrated Pollution Prevention and Control (IPPC) is implemented into UK law. This Directive (96/61/EC) was adopted on 24 September 1996 and must be implemented by 30 October 1999. The government is currently in the process of consultation on how it should be implemented.

Further controls on air pollution

Further and separate controls are imposed in respect of emissions to air under the Clean Air Act 1993, which consolidated the earlier Clean Air Acts of 1956 and 1968. Essentially, it aims to control emissions of smoke, dust, grit and fumes from all fires and furnaces. It includes provisions relating to:

- the height of chimneys;
- prohibitions and controls on certain emissions from certain premises;
- the requirement for new furnaces to be smokeless;
- the designation of 'smoke control areas' and the prohibition of emissions of smoke in those areas; and
- the obtaining of information on emissions from premises.

Generally, it is an offence to emit dark smoke from the chimney of any building, in particular from chimneys serving furnaces, boilers or industrial plant. It is also generally an offence to emit dark smoke from any industrial or trade premises. The requirements of the 1993 Act are distinguishable from other environmental legislation in that for the most part they do not require the granting of or compliance with a licence or permit.

The criminal sanctions for failure to comply with the provisions of the 1993 Act are set out in Table 3.

Waste management

Part II of the EPA significantly altered the regulation of waste disposal and treatment and the regulatory bodies with responsibility for such activities. It introduced a new waste management licensing regime which was brought into operation on 1 May 1994. Responsibility for the regulation of waste disposal has now been transferred to the Environment Agency.

Table 3. Maximum criminal sanctions for environmental offences

Offence	Summary (Magistrates' Court)	Indictment (more serious offences; Crown Court)
IPC/APC		
Failure to comply with certain IPC and APC obligations under Part I EPA 1990 (s23), e.g. operating a prescribed process without authorisation from Environment Agency	£20 000 and/or 3 months' imprisonment	Unlimited fine and/or 2 years' imprisonment
Air		
Failure to comply with prohibition on emission of dark smoke from chimney of any building under clean Air Act 1993 (s1)	£5000 (with £1000 maximum in respect of domestic premises)	n/a[a]
Failure to comply with prohibition on dark smoke from industrial premises under Clean Air Act 1993 (s2)	£20 000	n/a
Waste management		
Unauthorised or harmful deposit, treatment or disposal of waste on land (s33 EPA 1990)	£20 000 and/or 6 months' imprisonment	Unlimited fine and/or 2 years' imprisonment (5 years if special waste)
Failure to comply with Duty of Care as respects waste (s34 EPA 1990).	£5000	Unlimited fine
Failure to comply with special waste requirements (Regulation 18 Special Waste Regulations 1996)	£5000	Unlimited fine and/or 2 years' imprisonment

Table 3 continued

Offence	Summary (Magistrates' Court)	Indictment (more serious offences; Crown Court)
Water		
Causing or knowingly permitting pollution of controlled waters under WRA (s85)	£20 000 and/or 3 months' imprisonment	Unlimited fine and/or 2 years' imprisonment
Discharging trade effluent into public sewers without consent under WIA (s118)	£5000	Unlimited fine
Statutory nuisance (see below)		
Causing statutory nuisance under Part III EPA 1990 and failing to comply with abatement notice without reasonable excuse (s80)	£5000 plus £500 per day for continuing after conviction (£20 000 for noise from industrial, trade or business premises)	n/a

ª n/a, Not applicable.

   Section 33 of the EPA contains the broad requirement that a person who treats, keeps, deposits or disposes of controlled waste must have a waste management licence. To find the details of the regime, however, one must look to The Waste Management Licensing Regulations 1994 (SI 1994/1056). In addition, Section 33 creates several criminal offences which for the most part relate to complying with the waste management licensing system. It also provides that it is a criminal offence to treat, keep or dispose of controlled waste in a manner which is likely to cause pollution of the environment or harm to human health (Section 33(1)(c)).

   The 'duty of care as respects waste' (usually just called the 'Duty of Care' and so called in this chapter) under Section 34 of the EPA has been fully in force since 1 April 1992. It applies to anyone who imports, produces, carries, keeps, treats or disposes of waste, or as a broker has control of such waste. While applying to a wide range of people who may be regarded as holders of controlled waste (including industrial and commercial waste), occupiers of domestic property and the household waste they produce are specifically excluded from the scope of the Duty of Care.

Under the Duty of Care, holders of controlled waste must take all reasonable steps to:

- prevent any other person contravening Section 33 (i.e. the provisions relating to the treatment, keeping, deposit or disposal of controlled waste);
- prevent the escape of waste from their own control or that of another person;
- ensure that waste is only transferred to an authorised person or to a person for authorised transport purposes;
- ensure that waste when transferred is accompanied by a waste transfer note containing adequate written description of the waste.

These obligations are limited only to taking measures which are reasonable in the circumstances. Transferors and transferees of waste are required to keep records of waste held for at least two years. It is a criminal offence not to comply with the Duty of Care. The criminal sanctions for failure to comply with the Duty of Care are set out in Table 3.

Stricter controls are imposed on the producers, carriers and disposers of special waste under The Special Waste Regulations 1996 (SI 1996/972). A system of consignment notes is used to track special waste from the place where it is generated to its place of disposal. The producer, carrier and disposer must complete and retain copies of consignment notes for three years. Prenotification provisions also exist, stating that the office of the Environment Agency in the area of destination of the waste must receive copies of these notes. In addition, the disposer of special waste must keep site records which show deposits of special waste on a site plan. It is also stipulated by regulation that different categories of special waste or special waste and other waste must not be mixed. The failure of a producer, disposer or carrier of special waste to comply with the regulations is a criminal offence. The criminal sanctions for failure to comply with these requirements are set out in Table 3.

Water Resources Act 1991 ('WRA')

The WRA provides for a system of consents for discharges into controlled waters, widely defined to include any groundwaters, inland freshwaters, coastal waters and territorial sea waters (Section 104(1)). The consents are known as 'discharge consents' and after 1 April 1996 they became the responsibility of the Environment Agency. Although discharge consents may be unconditional, more usually conditions are attached setting emission limits for specific substances or relating to the quality of the discharge. Conditions may also require that there be monitoring, treatment of discharges and record keeping.

Unlike other environmental discharges there is no requirement to have a permit in order to be able to discharge into controlled waters. Accordingly, it is not an offence to make a discharge into controlled waters without a discharge consent. Instead, the existence of a discharge consent operates as a defence to Section 85 of the WRA, which prescribes the offences of causing or knowingly permitting any poisonous, noxious or polluting matter or any solid waste to enter into controlled water. The criminal sanction for breaching Section 85 is set out in Table 3.

Under Section 161 of the WRA 1991 the Environment Agency has the power to carry out remedial or restoration work if it is necessary to prevent or remedy pollution. This power is likely to be supplemented soon by additional powers under Section 161A-D of the WRA, which was inserted by the Environment Act 1995, but which has not yet come into force. It will empower the Agency to serve a works notice upon a person who has caused or knowingly permitted pollution to take place, requiring that person to carry out appropriate remedial or restoration work. The WRA also contains enforcement powers such as the power to enter premises, to request information and to institute criminal proceedings.

Water Industry Act 1991 ('WIA')

Discharges of trade effluent into the public sewers are governed by the WIA. A consent must be obtained or an agreement to discharge must be reached with the local sewerage undertaker. If the trade effluent contains one or more of the hazardous substances known as 'red list' substances, then the Environment Agency rather than the water company will determine whether or not to grant a consent and will decide what conditions to attach to it. Failure to have obtained a consent or breach of any conditions of a consent is a criminal offence as illustrated in Table 3.

Contaminated Land Regime

The basic provisions of the new contaminated land regime (the 'Contaminated Land Regime') were put on the statute book by Section 57 of the Environment Act 1995, but at the time of writing have still not been brought into force. A new Part IIA will be inserted into the EPA containing the principles and outline of the regime. This will be substantially supported by statutory guidance which is still in draft form.

It is likely that the delay in its implementation will be further extended until at least 1999.

The purpose of the Contaminated Land Regime is to deal with historical contamination of land. It will sit alongside the existing regimes, under which to a certain extent there may be clean-up of contaminated land, such as planning law, waste-management licensing, IPC and the statutory nuisance provisions. The Contaminated Land Regime makes provision for the identification and remediation of contaminated land in England, Scotland and Wales. It will be enforced primarily by local authorities and to a lesser extent, in England and Wales, by the Environment Agency, and in Scotland, by the Scottish Environment Protection Agency. In basic terms, each local authority will be under a duty to inspect its area from time to time to identify 'contaminated land'. Contaminated land is defined on a risk-based approach such that to be contaminated there must be significant harm resulting or a significant possibility of such harm or pollution of controlled waters. An integral part of the risk-based approach is the identification of a source–pathway– target relationship or 'pollution linkage'.

The enforcing authority, once it has identified such contaminated land, will then have the power to serve a remediation notice on an 'appropriate person' (or 'persons') requiring clean-up of contaminated land. Failure to comply will be a criminal offence backed by substantial fines and the power for the enforcing agency to do the remediation work itself and charge the 'appropriate person'. There are complicated provisions to determine who are 'appropriate persons', but they include the original polluter and, if the original polluter cannot be found, the present owner of the land.

Who may be liable under environmental statutes?

Prosecutions under environmental statutes such as those described above are frequently taken by the relevant regulatory authority against a company where a breach of environmental control has occurred in the course of a company's business. In principle, employees can also be held liable where they have been directly responsible for a breach of the law.

To facilitate prosecutions of the individuals responsible for running companies, all the environmental statutes (and others, such as those relating to health and safety) have a section, in what is now virtually a standard format, which provides:

> Where an offence . . . committed by a body corporate is proved to have been committed with the consent or connivance of, or to be attributable to any neglect on the part of, any director, manager,

secretary or other similar officer of the body corporate or any person who was purporting to act in any such capacity, he as well as the body corporate shall be guilty of that offence and shall be liable to be proceeded against and punished accordingly.

The above words are drafted widely enough so as to make a shareholder or parent company of a body corporate liable if acting as a shadow director. An example may be found in Section 157 of the EPA.

Civil Liability

Introduction

The main way in which civil liability may arise in the environmental context is under the law of tort or breach of contract. In this section the focus is on tort. There are a number of specific torts which may be relevant to environmental protection and these are discussed below.

Liability in tort arises where one person (the tortfeaser) commits a civil wrong against another person (the victim), so that the victim suffers damage or loss. No legal relationship between the victim and the tortfeasor is required. Instead, the law of tort sets a standard of behaviour that parties must observe, in default of which they must pay compensation or perform some other remedy.

In the environmental context, disputes between the owners or occupiers of neighbouring or adjacent land are normally based on one of four well-established causes of action in tort. These are:

- nuisance;
- the rule in *Rylands* v. *Fletcher*;
- negligence; and
- trespass.

They are discussed in turn below.

Nuisance

Traditionally, there have been two types of nuisance which are actionable in tort — private nuisance and public nuisance. In addition, the law of statutory nuisance, which may impose criminal liability, represents a third type.

Private nuisance is a cause of action which may be used by parties seeking compensation for damage to their interest in the environment. A nuisance arises where an act or omission on land unreasonably interferes with or disturbs a person's reasonable use or enjoyment of their land neighbouring or adjacent to the land from which the nuisance emanates. It is important to bear in mind, however, that the law of nuisance is directed towards compensating for interference with proprietary interests, rather than regulating the conduct of individuals, and so may not always be capable of providing a solution to environmental problems. Unlike proceedings in negligence, therefore, an action under the law of nuisance calls for the plaintiff to have a legal interest in the land affected.

Public nuisance is a criminal offence as well as a tort and only applies if the nuisance affects a significant part of the public as a whole. An example would be where contaminated land is polluting a source of drinking-water supply.

The third kind of nuisance is statutory nuisance, which may give rise to criminal rather than civil liability. Part III of the EPA sets out activities or occurrences which, provided they are 'prejudicial to health or a nuisance', will be a statutory nuisance; these include:

- the emission of smoke, fumes or gases from premises;
- the general (physical) state of premises;
- any dust, steam, smell or other effluvia arising on industrial, trade or business premises;
- any accumulation or deposit; and
- the emission of noise from premises.

If a local authority is satisfied that a statutory nuisance exists, it can serve an 'abatement notice' on the person causing it. The notice will, as the name suggests, require abatement of the nuisance and may prohibit or restrict its recurrence. The notice may also require the carrying-out of works necessary to achieve the aims of the abatement notice.

Action to abate a statutory nuisance may also be initiated by members of the public. Any individual who is aggrieved by a statutory nuisance can apply to the Magistrates' Court seeking an 'abatement order', which may contain the same requirements as an abatement notice relating to the abatement and the prevention of recurrence of the nuisance.

Where an abatement notice or an abatement order is obtained it will generally be served on the person responsible for the nuisance, i.e. the person to whose act or default the nuisance is attributed. Breach of or failure to comply with an abatement notice or order is a criminal offence with sanctions applying as set out in Table 3.

The rule in Rylands v. Fletcher

Rylands v. *Fletcher* was a case decided by the House of Lords in 1865, which has given its name to a tort concerning the escape of dangerous items or substances from land. The principle established in this case is that a person who brings anything onto his land for certain purposes is responsible for the foreseeable damage resulting from its escape. The rule also applies where a person allows accumulation on his land of something which escapes. One important restriction on the rule is that the substances kept on land that escape must have been kept there by way of 'non-natural user'. This has been interpreted by the courts to mean 'abnormal' rather than just 'artificial' use of land. The most common examples would include damage resulting from fuels and oils, toxic chemicals or anything that might cause physical damage, such as water contained in a tank. It is no defence to prove that all possible precautions were taken to prevent damage resulting from an escape, as the tort is one of strict liability.

The rule in *Rylands* v. *Fletcher* was recently considered at some length by the House of Lords in the landmark case of *Cambridge Water Company* v. *Eastern Counties Leather plc* ([1994] 1 All ER 53). This case is perhaps the most well-known recent case to deal with the issue of civil liability for contamination. It concerned the pollution of groundwater available for extraction for the public water supply at a borehole owned by the Cambridge Water Company. The pollution was caused by regular spillages of perchloroethene (PCE), which was used as a solvent at a leather tanning works owned by Eastern Counties Leather plc. The spillages ended in 1976, but the PCE had already seeped into the ground beneath the factory and migrated towards the borehole which was situated over a mile away. Some time after the spillages ended, the UK issued regulations to implement the 1980 EC Directive on drinking water (EC Directive 80/778 relating to the quality of water intended for human consumption) and water extracted from the borehole could not, because of the presence of PCE, be lawfully supplied as drinking water in accordance with those regulations. As a result, the Cambridge Water Company was forced to find a different source of supply at very considerable expense. The central feature in the decision was that foreseeability by the defendant of the type of damage complained of was a prerequisite for strict liability under the rule in *Rylands* v. *Fletcher* as well as for nuisance. The House of Lords held that it was necessary for such a requirement since it was not appropriate to impose retrospective liability for this kind of pollution, in the sense that spillages of PCE took place before the relevant legislation came into existence. It was established that as a matter of fact those responsible for the spillages at Eastern Counties Leather plc could not, at the time the spillages were made, have reasonably foreseen the damage to the borehole and so could not be held liable for it.

Negligence

In order to succeed in a negligence action a plaintiff must show some fault on the part of the defendant. To bring a claim in negligence the plaintiff must show that:

- a duty of care was owed by the defendant to the plaintiff;
- the defendant breached that duty by some act or omission to act; and
- there was foreseeable damage resulting from the breach.

Negligence has rarely been used in an environmental context and it is of limited utility in bringing an action for environmental protection.

Trespass

Trespass arises where there is a direct unauthorised interference with private property which is intentional or negligent. It is the immediate action itself which constitutes the wrong. It is sufficient only to show that the trespass has occurred and there is no need to show damage, which is a great advantage if one is trying to obtain an injunction. In the environmental context it is arguable that contaminants escaping from a contaminated site and entering upon another's land may constitute a trespass. A notable feature of trespass is that if a contaminant has moved from the defendant's land onto that of the plaintiff, the defendant is deemed to continue to commit an act of trespass for so long as that contaminant remains wrongfully on the plaintiff's land. In other words, a new trespass starts every day and accordingly, where it is available, an action in trespass can be valuable to a plaintiff.

Public registers of information

A recent development in environmental law is the requirement for regulatory authorities to maintain public registers of environmental information in their possession. This feature runs through each of the statutes mentioned earlier in the chapter, including the EPA, the WRA, the WIA, and the Radioactive Substances Act 1993. The main public registers held by the Environment Agency are shown in Table 4; in addition, there are some public registers of environmental information held by other regulatory bodies, such as local authorities.

The information recorded in each of these registers is available for public inspection. However, a number of the registers are subject to provisions which allow the exclusion of certain types of sensitive

Table 4. Public registers held by the Environment Agency

Register	Subject
Integrated Pollution Control (IPC) Register	Regulated industrial processes
Radioactive Substances Register	Use, build-up and disposal of radioactive materials
Water Quality and Pollution Control Register	Applications to put effluent into watercourses ('discharge contents'), standards for the quality of water (including beaches) and maps of defined coastal waters and freshwater limits
Water Abstraction and Impounding Register	Licence applications to take water from water courses or to store it for later use
Maps	Main rivers for each area covered by regional flood defence committees
Waste Management Licence Register	Waste and sites which manage waste
Carriers and Brokers of Controlled Waste Register	Applications to carry waste
Genetically Modified Organisms Register	Deliberate release into the environment
Chemical Release Inventory	Pollution released from industrial processes (including air, land and water) which are regulated.

information from the public register. Disclosure to the public of commercially confidential information such as trade secrets or details of secret manufacturing processes is prohibited. In addition, the Environmental Protection Act 1990 provides that no information shall be included in a register maintained under that Act where, in the opinion of the Secretary of State, the inclusion of such information in the register would be contrary to the interests of national security.

Public access to information on the environment has been increased over recent years, not least as a result of EC Directive 90/313 on the freedom of access to information on the environment. The Directive has been implemented in the UK by The Environmental Information Regulations 1992 (SI 1992/3240). Although this legislation will extend access to environmental information beyond merely that held in public registers, the effect of the Regulations in practice remains to be seen and will depend very much on the attitudes of local authorities.

3

European legislation on environmental assessment

Definition of environmental assessment

There is no legal definition of environmental assessment, but numerous publications on the subject, including for example guidance documents issued by the Department of the Environment, usually commence with a definition of the term. Accordingly, a common understanding has arisen of what environmental assessment comprises and involves. One such description of environmental assessment is:

> a technique for drawing together, in a systematic way, expert quantitative analysis and qualitative assessment of a project's environmental effects, and presenting the results in a way which enables the importance of the predicted effects, and the scope for modifying or mitigating them, to be properly evaluated by the relevant decision-making body before a decision is given. Environmental assessment techniques can help both developers and public authorities with environmental responsibilities to identify likely effects at an early stage and thus to improve the quality of both project planning and decision-making.
>
> (DoE Circular 15/88 on Environmental Assessment, 12 July 1988,
> paragraph 7 page 2)

This description identifies the characteristics and features of environmental assessment, which can be broadly represented by Figure 1.

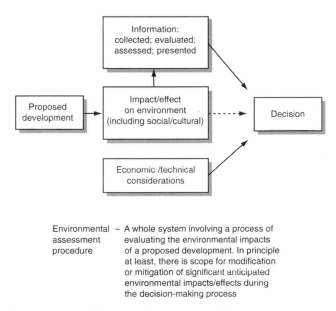

Environmental – A whole system involving a process of
assessment   evaluating the environmental impacts
procedure   of a proposed development. In principle
              at least, there is scope for modification
              or mitigation of significant anticipated
              environmental impacts/effects during
              the decision-making process

Fig. 1. Characteristics and features of environmental assessment.

The EA Directive

Introduction
A detailed analysis of the history and evolution of the policy and legislation on environmental assessment is beyond the scope of this book. Suffice it to say that in 1985, nearly five years after it had first been formally proposed by the Commission, the Directive on the assessment of the effects of certain public and private projects on the environment was adopted (the 'EA Directive') (Council Directive of 27 June 1985 on the assessment of the effects of certain public and private projects on the environment 85/337/EEC, OJ L175/40 of 5 July 1985). Prior to the EA Directive, not all member states had a system which provided for individual, case-by-case, scrutiny of the environmental impact of proposed development.

The EA Directive is one of the few pieces of European Community legislation which, in principle at least, embodies the preventative approach to environmental protection. As the preamble states:

> the best environmental policy consists in preventing the creation
> of pollution or nuisances at source rather than subsequently trying
> to counteract their effects

(EA Directive, Recital 1)

37

The EA Directive therefore addresses the need to take the effects of certain major projects on the environment into account at the earliest possible stage in, and ideally throughout, all technical planning and decision-making processes.

Broadly the philosophy behind environmental assessment is:

- to take account of concerns to protect human health;
- to contribute, by means of a better environment, to the quality of life;
- to ensure that the diversity of species is maintained; and
- to maintain the reproductive capacity of the ecosystem as a basic resource for life.
(EA Directive, Recital 11).

The EA Directive sets out certain common principles for environmental assessment and provides a framework for the implementation of procedures by member states for the purposes of achieving this. These procedures can either be integrated into existing decision-making procedures, or be the subject of entirely new ones (EA Directive, Article 2(2)). It must be remembered that, in itself, the EA Directive does not ensure the optimum and harmonised application of procedures across all member states. This is because in practice the implementation of the Directive and the enforcement of the national procedures for environmental assessment adopted under it are all left to the discretion of the member states. The EA Directive is project-orientated. The development of plans, policies and legislation and other decision-making tools are not subject to the EA Directive.

In March 1997 amendments were made to the EA Directive by Directive 97/11/EC (Council Directive 97/11/EC of 3 March 1997 amending Directive 85/337/EEC on the assessment of the effects of certain public and private projects on the environment, OJ L73/5 of 14.03.97). These amendments broaden the scope of projects subject to the requirement for environmental assessment, supplement the procedures involved and aim to improve the quality of the information provided as a part of the environmental assessment process. The amended provisions will have to be implemented by national legislation in the member states by 14 March 1999 at the latest. However, until national implementing legislation is brought into force, it is the unamended EA Directive which is relevant to applications for development. Therefore, this chapter deals mainly with the unamended EA Directive. The changes that will be brought about when Directive 97/11/EC is implemented by national legislation are discussed at the end of this chapter.

Aims of the EA Directive

The EA Directive provides a framework for implementing procedures for environmental assessment and requires that member states introduce measures to implement and meet the requirements set out in it. It is important to recognise that since the legislative instrument used is a Directive rather than a Regulation (for the distinction between these instruments, see Chapter 2) the method and detail of implementation are left to the discretion of each member state. As a consequence considerable variation has been experienced among member states in the methods and effectiveness of its implementation.

The aims and objectives of the EA Directive are as follows:

- to harmonise across all member states the principles of and laws relating to environmental assessment in order to avoid the creation of unfavourable competition conditions likely to affect the functioning of the common market (EA Directive, Recital 2);
- to introduce general principles for the assessment and evaluation of the effects on the environment of certain projects which are likely to have a major effect on the environment and to provide for the implementation of procedures to supplement and co-ordinate development consent procedures in order to achieve this (EA Directive, Recitals 1 and 5);
- to ensure that such assessment and evaluation is carried out prior to the grant of development consent for projects (EA Directive, Recital 6);
- to provide for the supply, primarily by the developer, of a minimum amount of appropriate information on a project and its effect on the environment, with scope for the provision of supplemental information from relevant authorities or others who might be concerned by the impact of a project on the environment (EA Directive, Recitals 6 and 10); and
- to provide information to the public concerning projects and their environmental effects.

Projects to which the EA Directive applies

The Directive applies to certain 'public and private projects which are likely to have a significant effect on the environment' (EA Directive, Article 1(1)). 'Project' is widely defined and means:

- the execution of construction works or other installations or schemes; and

- other interventions in the natural surroundings and landscape, including the extraction of mineral resources (EA Directive, Article 1(2)).

Hence the range of projects to which in principle the EA Directive applies is unclear and does not have any clear-cut boundaries. Potentially it covers all civil engineering projects, both private and public. However, the wording of Article 1(1) together with Article 2(1) narrows this down by stipulating that the projects which are to be the subject of an environmental assessment are those which are likely to have a significant effect on the environment by virtue of, among other things, the size, location or nature of the project in question. Furthermore, the EA Directive distinguishes between those projects which must always be the subject of an environmental assessment, as listed in Annex I of the EA Directive (Annex I projects), and those projects which must only be subject to an environmental assessment where member states consider that the characteristics of the project are such that an environmental assessment is required, as listed in Annex II of the EA Directive (Annex II projects) (EA Directive, Articles 4(1) and (2)). The full text of the EA Directive including Annex I and Annex II is set out in Appendix 1 of this book.

The words 'significant effect' are not defined by the EA Directive and therefore remain subjective and open to interpretation. However, in effect Annex I projects are presumed to have a significant effect since environmental assessment is mandatory in respect of them. For Annex II projects it has been left to the discretion of member states to decide the criteria and thresholds of significance which if met or exceeded would require that an environmental assessment be carried out. It is of note that early drafts of the proposed EA Directive included indications that the European Commission had intended that the EA Directive would contain certain criteria and guidelines which member states would have been bound to apply in determining which projects would be subject to assessment. However, in the final draft the dividing line between significant and insignificant effects is unclear and no guidance is offered to member states on where the dividing line should be.

There are several exceptions and exemptions to the general position (discussed above) on those projects which are subject to the requirement for environmental assessment:

- Projects which serve national defence purposes are excluded from the scope of the EA Directive (EA Directive, Article 1(4)). What will be regarded as a project for national defence purposes is not clear, although presumably it could be interpreted very widely to include, for example, large-scale boilers used to generate electricity to power a military establishment.

- Projects, the details of which are adopted by specific acts of national legislation, are also not covered by the EA Directive (EA Directive, Article 1(5)). The rationale for this (as expressed in the EA Directive) is that the Directive's objectives are met by the provision of information during the legislative process — including, one presumes, environmental information. It is doubtful, however, whether this is what happens in practice, and this exemption is a very broad one.

- Projects can be exempted by member states on an individual basis 'in exceptional cases' (EA Directive, Article 2(3)). Again, there is no indication as to what 'exceptional cases' may mean and the exemption merely sets out the procedural steps which apply to claim this exemption. Member states have full discretion to apply the exemption where appropriate, since the European Commission does not need to agree to the grant of an exemption. However, full information regarding any exemption granted by a member state must be given to the public and to the Commission. The Commission forwards this information to other member states and reports annually to the Council. However, statistics on the frequency and types of projects to which this exemption has been and is applied are not readily accessible, so it is difficult to ascertain its extent and effect in practice.

Annex I and Annex II projects

The main distinction between the way in which Annex I and Annex II projects are treated is that environmental assessment is mandatory for Annex I projects. These are generally large-scale projects, both public and private, which are deemed to have serious environmental effects (see Table 5). For Annex II projects, environmental assessment is only mandatory where the member states consider the characteristics of the proposed project are such that environmental assessment is necessary. The categories of Annex II projects are also listed in Table 5. A project which falls within Annex II will require an environmental assessment if found to meet or exceed the thresholds or criteria set by an individual member state for deciding whether the characteristics of a listed project are such that an environmental assessment is necessary. The discretion of member states to set the thresholds and criteria for deciding whether Annex II projects should entail an environmental assessment has led to considerable variation across the European Community as regards the

41

Table 5. Summary of Annex I and Annex II projects[a]

Annex I projects Mandatory environmental assessment	Annex II projects Environmental assessment only where characteristics so require
Oil refineries Large thermal power stations and nuclear power stations and reactors Installations for final storage or permanent disposal of radioactive waste Iron and steel works Installations for extracting and processing asbestos Integrated chemical installations Construction of motorways, express roads, railway lines and airports Trading ports and inland waterways Installations for incineration, treatment or landfill of hazardous waste	Generic projects are classed under the 12 headings: Agriculture Extractive industry Energy industry Processing of metals Manufacture of glass Chemical industry Food industry Textile, leather, wood and paper industries Rubber industry Infrastructure projects Other miscellaneous projects (e.g. leisure resorts, racing and test tracks for cars, storage of scrap iron) Modifications to projects included in Annex I for the development and testing of new methods or products for a period of up to one year.

[a] Appendix 1 gives a full list of Annex I projects and the Annex II classes of generic projects under the 12 headings in the table.

frequency and effectiveness of environmental assessment for listed Annex II projects. For example, in the United Kingdom the thresholds set are indicative and really for guidance only. Therefore the decision is based upon the specific circumstances of the project in question. By comparison, in the Netherlands the thresholds are absolute and inflexible. The decision as to whether or not a project listed in Annex II requires an environmental assessment will be taken by the relevant 'decision-making body' as designated in the member state's implementing legislation.

Impact on the environment
The general objective of environmental assessment is to identify, describe and assess in an appropriate manner both the direct and indirect effects

of an Annex I or Annex II project on the following components of the environment:

- human beings, fauna and flora (i.e. the living environment);
- soil, water, air, climate and the landscape (i.e. the non-living environment which supports or is essential to the living environment); and
- material assets and the cultural heritage;

and the interaction between the above three components (EA Directive, Article 3).

Although they are mostly self-explanatory, none of the above terms is defined. In effect, the meaning given to 'environment' is very wide and extends beyond merely the natural environment of air, water, land, plants and animals, to include the human and man-made environment and cultural heritage. Also of note is the geographical scale of the definition of environment, which extends to local, national, regional, continental and global impacts. For example, impact on 'climate' could include the global climate (for example, the potential for ozone depletion or global warming) as well as the micro or local climate.

Environmental assessment — the requirements

The person responsible for carrying out an environmental assessment is the developer. The developer is defined by the EA Directive as the applicant for authorisation for a private project or the public authority which initiates a public project (EA Directive, Article 1(2)). Therefore a public authority which initiates a project is, for this purpose, a developer.

The framework for environmental assessment as set out in the EA Directive can be divided into three main stages:

- Stage 1: information gathering and evaluation carried out by the developer. This includes the provision of such information on the project and its environmental impact to the decision-making body.
- Stage 2: consultation and provision of information from Stage 1. The decision-making body consults and provides information to the authorities with environmental responsibilities and third parties (for example the public and/or, in certain circumstances, other member states);
- Stage 3: the decision. The decision-making body considers all information which has been provided by the developer (Stage 1) and which arises during the consultation/provision of information process (Stage 2). This stage includes providing and making available

information regarding the decision taken to third parties, for example the public and/or, in certain circumstances, other member states.

The detail within this framework is a matter left to the discretion of each member state to pursue under its own national law.

Stage 1: information provided by the developer

The most prescriptive part of the EA Directive is that which identifies the information to be gathered and evaluated in respect of the environmental impact of the proposed project. This forms the core of the environmental assessment process. Member states are required to introduce measures under national laws to ensure that the developer supplies at least a minimum amount of such information. There are no preliminary 'scoping' requirements for the information to be gathered and evaluated, although the 1997 amendments to the EA Directive do introduce some 'scoping' provisions (described later in this chapter). In effect, the limits of such information are defined only by the general list of broad subject matter identified by the definition of environment.

The minimum information which the EA Directive requires (EA Directive, Article 5(2)) is:

- a description of the project, including information on the site, the design and size of the project;
- a description of the measures envisaged in order to avoid, reduce and, if possible, remedy any significant adverse environmental effects;
- the data required to identify and assess the main environmental effects which the project is likely to have on the environment; and
- a non-technical summary of the information derived from the above descriptions and data.

More detailed guidance on the information to be provided is set out in Annex III of the EA Directive; the full text of Annex III is given in Appendix 1. It includes, for example, information on alternatives, forecasting methods used to assess the effects on the environment, technical difficulties, and the indirect and secondary effects of the project on the environment. It supplements the minimum level of compulsory information specified in Article 5(2). The EA Directive is unclear as to the extent to which the provisions of this further Annex III information are compulsory. The reader of Annex III is left with no clear indication of the amount and detail of information which is to be provided.

The EA Directive does not specify the means by which the information is to be provided to the decision-making body. For example, the type of document to be produced following the information-gathering and evaluation phase is not stated. It is only stipulated that the provision of information shall be in an 'appropriate form' (EA Directive, Article 5(1)). In many member states, including the UK, the national implementing legislation provides that the information which is gathered and evaluated is to be formulated into an 'environmental statement'. The use of the term 'environmental statement' has become commonplace as a result of the implementation of the EA Directive, rather than as a requirement of the EA Directive itself.

Stage 2: consultations by the decision-making body

The EA Directive requires member states to make provision to allow the document provided by the developer pursuant to Stage 1 to be made available to certain categories of person. In addition some of these categories must also be given the opportunity to express their opinion on the application or request for development consent. The categories of persons to whom information must be made available are (EA Directive, Article 6):

- those authorities likely to be concerned by the proposed project by reason of their specific environmental responsibilities;
- the public; and
- in certain circumstances, other member states.

As regards consultation with authorities with specific environmental responsibilities, the EA Directive merely requires that the consultation occurs and no more (EA Directive, Article 6(1)). As regards consultation with the public, the EA Directive requires that the request for development consent and any information gathered under the EA Directive be made available to the public and that the public concerned must be given the opportunity to express their opinion before the project is initiated (EA Directive, Article 6(2)). However it is for member states to determine who are the 'public concerned' and decide the detail of the consultation arrangements (EA Directive, Article 6(3)). In this regard, the following practical arrangements are stipulated by the EA Directive:

- the specification of the place(s) where information can be consulted by the public;
- the specification of the way in which the public can be informed of the existence and availability of the information, e.g. by local newspaper advertisements, exhibitions and posters;
- the determination of the manner in which the opinion of the public will be secured, e.g. by written submissions or public enquiry;

- the setting of an appropriate time limit for consultation in order to ensure that decisions are made within a reasonable time.

As regards consultations with other member states, if a project planned for one member state is likely to have significant effects on the environment of a second member state (or if the latter makes a request) then the document provided by the developer pursuant to Stage 1 must be made available to that second member state (EA Directive, Article 7). The European Commission is of the view that where the environment of a neighbouring community may directly or indirectly be affected by a project, then it is right that the environmental assessment extends to cross-border impacts. In other words an environmental assessment should include all significant effects a project is likely to have on the environment, irrespective of the member state in which those effects are felt (Answer to Written Question No. 111/92 by Mr Karl Partsch and Mr Manfred Vohrer (LDR) to the Commission of the European Communities, 07.02.92, OJ C235/32 of 14.09.92).

The reference in EA Directive Article 6(2) to the project being 'initiated' is confusing since this could literally be interpreted to mean that the opinion of the public could be sought before the project is actually started, but after the grant of the development consent. This means that a decision could be taken on whether to grant consent before the consultation with the public takes place.

It is important to note that, as regards the consultations, the EA Directive preserves and respects national provisions (legal or administrative) for industrial and commercial secrecy and the safeguarding of the public interest (EA Directive, Article 10). This means that in certain commercial circumstances, information gathered under Stage 1 may be withheld and not provided to third parties in the way described above. The terms of this exception are potentially extremely wide and the EA Directive offers no clear indication of just how wide they may be.

Stage 3: the decision

The means by which the information provided by the developer (Stage 1) and that resulting from consultations (Stage 2) is taken into account during the decision-making process is not set out in the EA Directive. Further, the EA Directive simply provides that information gathered from the developer, the public and other member states 'must be taken into consideration' in the development consent procedure (EA Directive, Article 8). In many ways this stage can be regarded as the 'assessment' part of environmental assessment. What is assessed is the environmental information from a number of sources — namely the developer, the public and other consultees.

The provisions of the EA Directive impose essentially procedural rather than substantive obligations. For example, the EA Directive does not go so far as to require a decision-making body to refuse consent where a project is likely, according to the results of Stages 1 and 2, to have a damaging or significant impact on the environment. Likewise, there is no requirement on the decision-making body to impose mitigating or preventative measures if the environmental assessment shows that the project is highly likely to have a significant impact on the environment. The requirement is merely to take that environmental information gathered through the environmental assessment process into consideration when deciding whether or not to grant development consent. Lastly, since the EA Directive is essentially procedural, it does not allow the European Commission itself to decide whether a development project should be carried out, since this is a decision for the decision-making body of the relevant member state. The effect of this is that in practice, environmental assessment is a means by which the anticipated environmental impacts of a project are brought to the attention of the relevant decision-making body, so that it may make a more informed decision. The EA Directive goes no further than that.

Once a decision has been taken, the relevant decision-making body is obliged to inform the public of the content of the decision, any conditions attaching to it and the reasons and considerations on which the decision is based (EA Directive, Article 9). However, where a decision not to carry out an environmental assessment for Annex II projects has been taken, the EA Directive does not require that this decision and the reasons for it are made available to the public (Answer to Written Question No. 3099/92 by Wilfried Telkämper (V) to the Commission of the European Communities 14.12.92, OJ C350/3 of 29.12.93). This will be changed in the future as the 1997 amendments to the EA Directive require that the decision and reasoning be made public even where the decision is that environmental assessment is not required (see later in this chapter). Furthermore, the EA Directive is silent on the issue of post-project monitoring of actual environmental impacts, which some would argue is, or at least should be, fundamental to the environmental assessment procedure (Sheate, 1994, p. 104).

The EA Directive in practice

The Review

Member states were allowed three years, until 3 July 1988, to comply with requirements of the EA Directive. In other words, they were required to take the necessary measures, whether by new or amended legislation and/or procedures, to meet the requirements of the EA Directive within this period. In theory, by the date of publication of this book there should have been a decade of experience and practice of the application and effect of the EA Directive. However, this has not been the case. In 1991, belatedly, but in accordance with a requirement of the EA Directive itself (Article 11(3)) the European Commission carried out a comprehensive review across all member states of the application and effectiveness of the EA Directive (the 'Review') for the six-year period from July 1985 to July 1991 (Report from the Commission of the Implementation of Directive 85/337/EEC on the assessment of the effects of certain public projects on the environment and Annex for the United Kingdom, Com (93) 28 Final — Vol. 12, 2 April 1993).

The Review found that although in many member states the implementation of the EA Directive had been slow and was consequently in its early stages, nevertheless there was evidence to show that the planning, design and authorisation of projects was, in 1992, beginning to be influenced by environmental assessment procedures and that the environment was benefiting. However, the Review recognised that the full potential of environmental assessment, in terms of environmental protection, was not yet being fully realised.

The results of the Review indicated that numerous difficulties had been encountered in the transposition of the requirements of the EA Directive into the national laws of member states. Several member states did not adopt provisions to implement the requirements of the EA Directive by 3 July 1988 and in many of those cases implementation was still not fully effective by July 1991. In addition, there had been a widespread failure to implement the requirements of the EA Directive in full. As a consequence, infringement proceedings have been brought or threatened by the European Commission against the majority of member states. For example, the European Commission commenced proceedings against the UK Government over the alleged failure to satisfy environmental assessment requirements for major infrastructure projects, including the East London River Crossing and the extension of the M3 motorway through Twyford Down, Winchester.

The Review attributed these problems to a number of factors:

- difficulties experienced by some member states (for example Italy) in implementing the EA Directive as a result of several different tiers of government, which increase the complexity of introducing the provisions of the EA Directive into existing systems of development control;
- the problem of securing the co-operation of and compliance by a number of different ministries and consequently in some cases the passing of a number of different legislative instruments in order to implement environmental assessment procedures;
- resistance of member states to consultation with the public and environmental authorities where this was in effect a new requirement and there was no prior established practice or legal requirement of this nature;
- difficulties with interpretation and meaning of words and terms used in the EA Directive, for example 'significant on the environment effect' (see above).

The introduction of environmental assessment into national administrative and legal procedures has for many member states proved a demanding exercise.

The Review identified the following shortcomings in the transposition of and implementation of the EA Directive into national laws:

- failure to include provisions in national laws for the assessment of the impact of entire classes of Annex II projects;
- failure to take the measures necessary to ensure that the developer provides all of the minimum specified information required pursuant to Article 5(2);
- failure to take the measures necessary for the obligatory provision of information to and consultation with the public;
- failure to take the measures necessary to consult neighbouring member states where a project has cross-border implications;
- failure to take the information gathered and evaluated during Stage 1 into consideration in the decision-making process (Stage 3);
- failure to publish the decision taken.

The Review also considered the experience of member states beyond the formal legal position of the EA Directive, in the application of environmental assessment in practice. While the total number of environmental assessments undertaken within the European Community was significant and increasing, the Review found that there were considerable variations between member states in the number of environmental assessments carried out and consequently in the coverage of projects likely to give rise to significant environmental impacts. On the basis of the limited experience obtained so far, these variations are thought to be reflections of the following factors:

- variations between those member states which had existing systems for environmental assessment and those member states without such systems;
- variations between those member states which have implemented the EA Directive by modifying existing procedures and those which have provided a separate and new system of environmental assessment;
- variations between member states in the quality and coverage of environmental statements.

The Review revealed that in several member states the majority of environmental statements were not of satisfactory quality. This is thought to be due to failure to commence environmental assessment at a sufficiently early stage in the development decision-making process, and a lack of sufficiently experienced staff. The resultant effect is that scope for the possible adoption of alternatives, either for the individual project or for its location or route, is rarely seriously considered since there is insufficient time to do so.

On a more positive note, the Review clearly established that the EA Directive has had beneficial effects in protecting the environment of member states, for example by:

- providing authorities with information on the environment to be used in the environmental assessment of individual project proposals;
- identifying ways of mitigating the impact of a project on the environment and to modify the project proposal in advance of project commencement;
- the formal involvement of environmental authorities in the decision-making process and project analysis, thus leading to a greater awareness of the impact of projects on significant biotopes in the European Community.

Overall there is evidence that the quality of environmental assessments is steadily increasing as experience and understanding of environmental assessment procedures and environmental impact develops and the number of practitioners experienced and qualified to carry out environmental assessments increases.

The amendments made to the EA Directive by Directive 97/11/EC are aimed at dealing with some of these problems and issues. A summary of the revisions to the EA Directive is given at the end of this chapter. The development of the environmental assessment practices and procedures has come a long way since 1988. However, as the Review and revisions to the EA Directive in 1997 tend to indicate, there is still room for improvement and further development of environmental assessment methodologies and practices.

The EA Directive and the doctrine of direct effect

As explained in Chapter 2, the European Commission has traditionally, and until recently always, used Directives rather than Regulations as the legal instruments for European Community legislation on protection of the environment. Whereas Regulations apply directly to the individuals or other legal entities within member states without the need for any implementing legislation, Directives require national laws to implement them. Accordingly, generally speaking the requirements of a Directive are not directly applicable in member states and cannot place legal obligations upon individuals unless the Directive in question has been implemented by means of national legislation. However, where the required national legislation has not been brought into place by the implementation date stated in a Directive, or where the national legislation does not give full effect to a Directive, the doctrine of 'direct effect' may apply.

The doctrine of direct effect has been developed from a long line of cases decided by both the ECJ and national courts of member states. It is only possible to give a brief summary of the doctrine in this book. Its effect is that, subject to satisfying certain requirements, a Directive can have direct legal effect in member states after its date for implementation has expired, if the member state has failed to transpose the directive fully into its national legislation. The direct effect doctrine only applies if the provisions of the Directive in question are sufficiently clear and precise, unconditional and leave no room for discretion in implementation. The basis of the doctrine is that a member state should not be allowed to avoid liability by reason of its own failure to transpose a Directive either in time or with full effect. As a consequence of the doctrine, it is possible for individuals to bring proceedings in their national courts to enforce directly the provisions or requirements of a Directive, but only against the government of the member state, or governmental body. This is subject to the requirement that:

- the date for implementation of the Directive has passed;
- the Directive is unconditional and sufficiently clear and precise in its terms to grant rights to an individual; and
- the rights of the individual have been infringed and damage has been suffered as a result of the failure to transpose the Directive.

The rights of an individual against the government of a member state also extend to any 'emanation' of that government. This has been interpreted by the ECJ as follows (Case C-188/89 *Foster and Others v. British Gas plc* [1990] 3 All E.R. 897):

> . . . a body, whatever its legal form, which has been made responsible, pursuant to a measure adopted by the state, for providing a public service under the control of the state and has for that purpose special powers beyond those which result from the normal rules applicable in relations between individuals, is included in any event among the bodies against which the provisions of a Directive capable of having direct effect may be relied on.

Although the courts have yet to give a definitive ruling on the application of the doctrine of direct effect to the EA Directive, the point has been discussed in a number of cases, leaving open the possibility of an application on this basis. There have been two types of circumstances in which the courts have considered the direct effect of the EA Directive. In some cases, the applicant has pleaded that the EA Directive has not been properly implemented by the EA Regulations and so has sought to rely directly on the provisions of the EA Directive. In others, sometimes referred to as 'pipeline cases' (as the projects concerned were 'in the pipeline' when the UK legislation was brought into force), the planning application was submitted between the prescribed date for implementation of the EA Directive and the date of actual implementation in the UK (i.e. the date when the EA Regulations were actually brought into force, which was nearly two weeks later). In the latter instance, applicants have been forced to plead the direct effect of the EA Directive as there were no relevant UK regulations in force at the time of the application. These cases should be distinguished from other 'pipeline' cases which involved projects for which the planning applications were made before the prescribed date for implementation, only to obtain approval afterwards. The courts have consistently stated that in these cases the EA Directive could not have been directly effective as the date of the planning application was prior to the date by which the member states were obliged to implement the Directive.

In *Twyford Parish Council and Others v. Secretary of State for the*

Environment ([1992] 1 Env LR 37) it was held that the EA Directive had direct effect in relation to Annex I projects. However, this part of the ruling is not strictly binding as, because of the timing of the relevant authorisations, the court held that the Directive did not apply at all to the case. By contrast, in the *Petition of Kincardine and Deeside District Council* ([1992] 1 Env LR 151) the Scottish Court of Session found that neither Article 2 nor Article 4 of the EA Directive could be read as imposing a precise and unconditional obligation which could be translated into direct effect. This was based on the fact that in relation to Annex II projects there is a discretion as to whether, in any particular case, an environmental assessment should be carried out, which in turn meant that there was insufficient precision and certainty in the provisions of the EA Directive for them to have direct effect with respect to Annex II projects. However, again this part of the ruling is not strictly binding, as (due to the question of dates and timing) it was held that the EA Directive did not apply. While on the face of it the *Twyford* and *Kincardine* cases may appear to contradict each other, that is not the case, since it is arguable that the *Twyford* case provides that the direct effect doctrine applies in respect of Annex I projects while the *Kincardine* case provides that the direct effect doctrine does not apply in respect of Annex II projects. Unfortunately, some confusion has been created by the decision in *Wychavon District Council* v. *Secretary for the Environment and Velcourt Limited* ([1994] Env LR 239) in which the High Court found that a Directive could only create a direct effect if every provision within it was capable of being directly effective, and therefore the EA Directive did not have direct effect. However, it is submitted that this approach is not substantiated by any ECJ case law on the direct effect doctrine. ECJ case law, on the contrary, indicates that each provision of a Directive may be taken separately and may or may not operate to create a direct effect. In the recent 'Dutch dykes' case, the ECJ, while managing to avoid answering precisely whether the EA Directive had direct effect, appeared to imply that certain provisions did indeed have direct effect. Therefore the *Wychavon* judgment may have proceeded on a misapprehension in this regard. The High Court in *R* v. *North Yorkshire County Council ex parte Brown* ([1996] NPC 160) followed the reasoning in *Wychavon* with regard to direct effect, but the decision did not turn on this issue. The judge accepted that the issue of direct effect had a wider significance than the facts before him, and gave leave to appeal to the Court of Appeal.

In summary, the case law illustrates that to the extent that they have not been fully implemented by the UK legislation, provisions of the EA

Directive may have a direct binding effect on local planning authorities. Although the Courts have not confirmed that provisions of the EA Directive can be relied on as directly effective, the opportunity is still there for them to do so. Overall this means that the importance of the EA Directive is enhanced in UK law in that the requirements and objectives of the EA Directive itself, as well as the national legislation enacted to implement it, must be considered by relevant decision-making bodies when they make development consent decisions.

Compliance in the UK

Role of environmental assessment within the town and country planning system
In the UK, the powers, duties and functions in relation to environmental protection are spread over a variety of organisations, some exercising policy-making functions and others regulatory functions. As a result there has been confusion over the jurisdiction of the different bodies involved. Nowhere is this more acute than in the case of environmental assessment.

In most cases environmental assessment is an integral part of the town and country planning system of development control. During the course of its negotiations, largely as a result of representations on the part of the UK government, the EA Directive was amended so as to resemble very closely the system of development control already in place in the UK (the town and country planning system). Indeed, the prevailing view of the UK government during the early stages of the negotiations over the EA Directive was that since local planning authorities were already, as a matter of law, always required to have regard to all material considerations during the granting of planning consent, in practice formal environmental assessment was unnecessary. This was because the general provision in effect meant that if significant, the environmental impact of the development or project was in any event taken into account during the decision-making process.

It is no surprise, therefore, that the transposition of the EA Directive into the national laws of the UK was largely within the existing town and country planning system. The approach taken by the UK government to meet the requirements of the EA Directive was that in practice it should not significantly add to existing procedures for authorising developments and projects. Therefore, environmental assessment was not intended to be imposed unless it was required by the EA Directive (Department of Environment Circular 15/88, paragraph 8, p. 2). With

54

one small exception, no primary legislation, i.e. Act of Parliament, was amended or adopted specifically for the purposes of implementing the EA Directive.

Interrelationship with other pollution control systems

In spite of being complementary to each other, the town and country planning systems and the system of pollution control under environmental law have developed independently and remain separate from each other. Both systems have a common aim in the protection of the environment, but both seek to achieve this aim from different perspectives and by different means. On the one hand, the town and country planning system regulates the location of development and the operations that take place in order to avoid or mitigate adverse effects of the use of land on the environment (including control after a development or main use of land has ceased). On the other hand, the pollution control system regulates the characteristics of potentially polluting activities and operations, for example by regulating emissions from industrial processes to air, water and land. The two systems were summarised in Government policy on the environment as follows:

> Planning control is primarily concerned with the type and location of new development and changes of use. Once broad land uses have been sanctioned by the planning process, it is the job of pollution control to limit the adverse effects that operations may have on the environment. But in practice there is common ground. In considering whether to grant planning permission for a particular development, a local authority must consider all the effects, including potential pollution; permission should not be granted if that might expose people to danger. And a change in an industrial process may well require planning permission as well as approval under environmental protection legislation.
>
> (This Common Inheritance, Britain's Environmental Strategy Command Paper CM 1200, paragraph 6.39, HMSO 1991)

Despite coming within the town and country planning system, some of the basic principles of environmental law are particularly appropriate to environmental assessment. One such principle is the precautionary principle which provides that even when the (exact) effects of a potentially harmful emission or discharge into the environment are not scientifically known or proven, it is presumed that the release of such substances should be prohibited. This is of particular importance in the context of environmental assessment.

55

Inevitably, increasing public awareness and involvement in environmental issues has meant that local planning authorities have increasingly become involved in and taken a greater interest in controlling potentially polluting activities. At the same time the pollution control system and environmental law generally have increased in scope and stringency and the pollution control authorities have been more willing to enforce such laws. Consequently it has been recognised that the relationship between the controls exercised under the town and country planning system and the pollution control system must not give rise to unnecessary duplication and conflict of interest between local planning authorities and pollution control authorities. It therefore follows that each of these authorities should be encouraged to consult closely with the other, but not to go so far as to duplicate controls which are the statutory responsibility of the other. The relationship between the town and country planning system and the environmental protection system has been the subject of case law and guidance issued by the Department of the Environment specifically on this topic. This highlights the difficulties in practice of deciding on the boundary between the two systems, in particular in the context of environmental assessment where the two systems are in close proximity.

The relationship and conflict between powers and duties exercised under the Town and Country Planning Act 1990 (i.e. the town and country planning system) and the powers and duties under Part I of the Environmental Protection Act 1990 (i.e. part of the pollution control system) was considered in *Gateshead MBC* v. *Secretary of State for the Environment and Northumbrian Water Group plc* [1994] (71 P.& C.R. 350, C.A.). This case concerned a planning application for a clinical waste incinerator in Gateshead which had obvious pollution control implications. In an appeal by the planning authority against the Secretary of State's decision to grant planning permission, the Court of Appeal outlined the relationship between the town and country planning system and the pollution control system. This can be summarised in the following terms:

- the starting point is the obligation to have regard to the development plan and other material considerations in granting the planning permission;
- the environmental impact of emissions to the atmosphere is a material consideration at the planning stage;
- it is not lawful to have a policy of hiving off the consideration of all environmental effects in their entirety, to be dealt with under the

pollution control system, which in this case was Part I of the Environmental Protection Act 1990;

- just as the environmental impact is a material planning consideration, so too is the existence of a stringent regime of pollution control for preventing or mitigating any environmental impacts and in this case rendering any emissions from the incinerator harmless;

- where two statutory controls overlap it is not helpful to attempt to define where one ends and another begins. To do so means that there is a danger of losing sight of the obligation to consider each case on its individual merits.

The Department of the Environment has issued a Planning Policy Guidance Note, 'Planning and Pollution Control' ('PPG23') which considers the relationship between the planning system and pollution control legislation. It is only guidance and, in practice, it will be for the courts to deal with issues of law. It was in draft form at the time of the *Gateshead* case and publication of the final version was actually delayed to incorporate the implications of that judgement. PPG23 incorporates the reasoning of the Court of Appeal in *Gateshead* and emphasises that the two regulatory systems of planning control legislation and pollution control legislation are separate but complementary, in that both are concerned with protecting the environment. PPG23 also advises that planning authorities should not seek to duplicate controls which are the statutory responsibility of other bodies and in particular that planning controls are not appropriate for the regulation of the detailed characteristics of potentially polluting activities. Planning authorities should assume that the relevant pollution controls will be appropriately applied and enforced and should not substitute their own judgement for that of pollution control authorities on such issues. However, PPG23 notes that there is no clear dividing line between planning and pollution control and that some matters may be relevant to both systems.

In summary, although the environmental impact of a planning proposal is a material consideration, it is for the local planning authority to decide in any one case, as a matter of planning law and judgement, whether potential pollution issues or problems arising from a development can be left to the control of the pollution control system and the relevant pollution control authority or whether the evidence of environmental issues or problems is so weighty that planning permission must be refused. In other words, there is likely to be little merit in attempting to resolve severe problems through the pollution control process. The *Gateshead* case and PPG23 are valuable indications of how the courts are likely to view and interpret the relationship between the town and country planning system and the pollution control system.

Implementing legislation for the EA Directive in the UK

The implementation of the EA Directive in England and Wales has occurred through a series of regulations (totalling at least 20), most of which are within the town and country planning system of development control. There are separate but similar regulations for Scotland and Northern Ireland. However, the detail of these is beyond the scope of this book. All references in this book to 'UK law' or 'environmental assessment in the UK' should be taken as relating only to England and Wales. If information on environmental assessment in Scotland or Northern Ireland is needed, then the specific provisions relating to those jurisdictions should be consulted. The regulations for England and Wales and Northern Ireland are listed in Table 6.

Broadly speaking, the legislation can be categorised into two types:

- Regulations for environmental assessment where planning consent is required within the normal town and country planning system and procedures. In essence this is where the development consent is to be granted pursuant to and in accordance with the Town and Country Planning Act 1990. The Town and Country Planning (Assessment of Environmental Effects) Regulations 1988 are the principal set of applicable regulations in England and Wales and are the focus of Chapter 4. These and associate regulations apply to the majority of Annex I and Annex II projects and account for around 70% of the total number of environmental assessments carried out and environmental statements produced in the UK each year (Department of the Environment, Transport and the Regions first consultation paper on the implementation of EC Directive (97/11/EC) on environmental assessment, 28 July 1997);
- Regulations for environmental assessment for projects which do not require planning consent pursuant to the Town and Country Planning Act 1990. A brief description of these regulations is set out at the end of Chapter 4. These projects can be divided into two sub-categories:
  - Those projects which fall within the town and country planning system, but which do not require planning permission since there is a deemed grant of planning permission. For example, in respect of many large infrastructure projects such as railway lines, trunk roads, ports and harbours, the authority to grant the

Table 6. Regulations implementing the EA Directive in England and Wales

Town and country planning system
The Town and Country Planning (Assessment of Environmental Effects)
Regulations 1988 (SI No. 1199) *as amended by*:

> The Town and Country Planning (Assessment of Environmental Effects)
> (Amendment) Regulations 1990 (SI No. 367)
>
> The Town and Country Planning (Assessment of Environmental Effects)
> (Amendments) Regulations 1992 (SI No. 1494)
>
> The Town and Country Planning (Assessment of Environmental Effects)
> (Amendment) Regulations 1994 (SI No. 677)

The Town and Country Planning General Development Order 1988 (SI No. 1813)
as amended by:

> The Town and Country Planning General Development (Amendment) Order
> 1994 (SI No. 678)

The Town and Country Planning (Simplified Planning Zones) Regulations 1992
(SI No. 2414)

The Town and Country Planning (Environmental Assessment and Permitted
Development) Regulations 1995 (SI No. 417)

The Town and Country Planning (Environmental Assessment and Unauthorised
Development) Regulations 1995 (SI No. 2258)

The Town and Country Planning (General Permitted Development) Order 1995
(SI No. 418)

The Town and Country Planning (General Development Procedure) Order 1995
(SI No. 419)

Afforestation
The Environmental Assessment (Afforestation) Regulations 1988 (SI No. 1207)

Electricity and pipeline works
The Electricity and Pipeline Works (Assessment of Environmental Effects)
Regulations 1990 (SI No. 442) *as amended by*:

> The Electricity and Pipeline Works (Assessment of Environmental Effects)
> (Amendment) Regulations 1996 (SI No. 422)
>
> The Electricity and Pipeline Works (Assessment of Environmental Effects)
> (Amendment) Regulations 1997 (SI No. 629)

Table 6 continued

Harbour works

The Harbour Works (Assessment of Environmental Effects) Regulations 1988 (SI No. 1336)

The Harbour Works (Assessment of Environmental Effects) (No. 2) Regulations 1989 (SI No. 424)

The Harbour Works (Assessment of Environmental Effects) Regulations 1992 (SI No. 1421) *as amended by*:

    The Harbour Works (Assessment of Environmental Effects) Regulations (Amendment) Regulations 1996 (SI No. 1946)

Salmon farming

The Environmental Assessment (Salmon Farming in Marine Waters) Regulations 1988 (SI No. 1218)

Roads

The Highways (Assessment of Environmental Effects) Regulations 1988 (SI No. 1241)

The Highways (Assessment of Environmental Effects) Regulations 1994 (SI No. 1002)

Land drainage

The Land Drainage Improvement Works (Assessment of Environmental Effects) Regulations 1988 (SI No. 1217) *as amended by*:

    Land Drainage Improvement Works (Assessment of Environmental Effects) (Amendment) Regulations 1996 (SI 1995/2195)

Transport works

The Transport and Works (Applications and Objections Procedure) Rules 1992 (SI No. 2902)

The Transport and Works (Assessment of Environmental Effects) Regulations 1995 (SI No. 1541)

---

development consent rests with the relevant Secretary of State rather than with a local planning authority;

o Those projects which do not fall within the town and country planning system and in respect of which approval for development falls under different legislation (i.e. not the Town and Country Planning Act 1990).

1997 Amendments to the EA Directive

Introduction

It has now been over ten years since the original EA Directive was adopted and, as might be expected, some of the provisions have recently been amended. On 3 March 1997 Directive 97/11/EC was adopted, which makes some significant changes to the original EA Directive. (Directive 97/11/EC will hereafter be referred to as the '1997 Directive' and the new consolidated version of the EA Directive as amended by Directive 97/11/EC will be referred to as the 'Amended EA Directive'. The original EA Directive (85/337/EC) will continue to be referred to simply as the 'EA Directive'.)

The main aims of the amendments introduced by the 1997 Directive are to clarify and supplement the EA Directive in order to ensure a more even application across all member states and to improve the quality and scope of the information provided as part of the environmental assessment process.

The provisions of the Amended EA Directive will not apply to applications for development consent in the UK until it is implemented by national legislation. The implementation date is not until 14 March 1999, so it cannot have any direct effect until after that date. Therefore, until implementing regulations are brought into force in the UK it is the old EA Directive that will be relevant. The UK government is in the process of drafting legislation to implement the changes. A consultation paper on the form of the implementing legislation for the 1997 Directive was published in July 1997 (the 'First Consultation Paper') and a further consultation paper focusing on the proposed 'indicative' and 'exclusive' thresholds was published in January 1998 (the 'Second Consultation Paper'). The following paragraphs outline the effects of the amendments to the EA Directive and discuss how these will affect the UK environmental assessment process when they are implemented, based on the government's proposals set out in the two consultation papers.

The main effects of the Amendments

The main effects of the amendments are:
- Increased coverage
  o Annex I and Annex II are expanded, so more projects will require environmental assessment;

- Changes to procedures
  - clarification of the way in which member states can decide whether Annex II projects require environmental assessment;
  - requirement for competent authorities to publicise their decisions on whether environmental assessment is needed *or not needed*;
  - increased scoping — informal scoping is to be encouraged and competent authorities, if asked by a developer, will have to give advice on the content of any proposed environmental statement;
  - requirement for competent authorities to give reasons for development consent decisions for all projects which have been subject to environmental assessment;
  - enhanced arrangements for consultation with other member states on projects likely to have significant transboundary environmental effects;
  - possible linkage with IPPC procedures.

Other changes have been made to the EA Directive, but these are mostly aimed to bring the procedures for environmental assessment in other member states more into line with the approach already taken in the UK. These amendments will therefore have little or no practical impact on the environmental assessment process in the UK.

The full text of the 1997 Directive is set out in Appendix 4 of this book.

Increased coverage for environmental assessment
The 1997 Directive adds 12 new classes of project to Annex I and six of the existing classes have been extended so as to include new projects. In fact, ten of the 'new' classes of project have simply been transferred from the original Annex II, thus making them always subject to the requirement for environmental assessment.

The 1997 Directive adds eight new classes of project to Annex II and ten of the existing classes have been extended. Many of the other classes have been clarified, consolidated or reorganised. The 1997 Directive also makes it clear that modifications to Annex II projects (as well as to Annex I projects) are considered to be Annex II projects in their own right. A summary of the most significant additions to the classes of projects in Annex I and Annex II as set out in the First Consultation Paper is given in Table 7. Appendix 4 gives full details of the new Annex I and Annex II.

Table 7. Summary of significant changes to Annex I and Annex II projects

	New project classes	Extensions to existing project classes
Annex I	Incineration or chemical treatment of non-hazardous waste (over 100 t/day)	Decommissioning added to nuclear power stations and other nuclear reactors
	Groundwater abstraction or artificial recharging ($10^7 m^3$/year or more)	Reprocessing and production of nuclear fuel added to storage/disposal of radioactive waste
	Transfer of water between river basins, above certain thresholds	Production of non-ferrous crude metals added to production of cast iron and steel
	Waste-water treatment plants (capacity of over 150 000 population equivalent)	New, wider definition of what constitutes an integrated chemical installation
	Extraction of petroleum (over 500 t/day) and natural gas (over 500 000 m^3/day)	Construction (or widening) of 10 km or more of dual carriageway
	Dams and other installations (holding back more than $10^7 m^3$)	Piers outside ports added to the definition of trading ports
	Oil, gas and chemical pipelines (over 40 km long and over 800 mm in diameter)	
	Intensive rearing of poultry or pigs (over certain thresholds)	
	Industrial plants for the production of paper (over 200 t/day) or pulp	
	Quarries and open-cast mining (over 25 ha) and peat extraction (over 150 ha)	
	Overhead power lines with a voltage of 220 kV or more (longer than 15 km)	
	Storage of petroleum, petrochemical or chemical products (200 000 t or more)	

Table 7 continued

	New project classes	Extensions to existing project classes
Annex II	Wind farms	Irrigation projects added to water management projects for agriculture
	Manufacture of ceramic products by burning	
	Coastal and maritime works capable of altering the coast	Forestry projects now include deforestation
	Groundwater abstraction and article recharging (below Annex I threshold)	All intensive livestock installations, not just those for pigs or poultry
	Transfer of water resources between river basins (below Annex I threshold)	Intensive farming of any fish, not just salmon
	Transfer of water resources between river basins (below Annex I threshold)	Construction of railways and intermodal transshipment facilities, and intermodal terminals (projects not included in Annex I)
	Installations for the recovery or destruction of explosives	Construction of port installations added to construction of harbours etc.
	Permanent camp sites and caravan sites	Installations for the disposal of all types of waste, not just industrial or domestic
	Theme parks	Ski-runs and other associated developments added to skiing developments
		All marinas now covered, not just yacht marinas
		Associated developments added to holiday villages and hotel complexes

Changes to the procedures for environmental assessment

Indicative thresholds and criteria

The 1997 Directive clarifies the way in which member states may determine whether Annex II projects require environmental assessment. Under the new Article 4(2), member states may now use thresholds or criteria; or consider projects on a case-by-case basis; or use a combination of both. Recital 8 of the preamble makes it clear that member states may set thresholds below which it will not be necessary to consider whether a project requires environmental assessment. A new Annex III sets out the range of criteria which must be taken into account in determining whether Annex II projects require environmental assessment.

To date, the UK has relied on a case-by-case approach to determine whether Annex II projects require environmental assessment, using DoE general guidance on the types of project likely to require environmental assessment and a set of published indicative criteria (see Chapter 4). The 1997 amendments to the EA Directive do not mean that this approach has to be fundamentally changed, but they do put forward options which are likely to improve the system. The Second Consultation Paper proposes that the case-by-case approach will continue to be the basis for determining whether Annex II projects need environmental assessment, but it will be supplemented by the use of new 'exclusive' thresholds and a revised list of 'indicative thresholds'.

The fundamental test will continue to be whether the proposed development is likely to give rise to significant environmental effects. In order to determine this question, the relevant competent authority will be able to use the selection criteria set out in the new Annex III to the Amended Directive, and also the existing government guidance. However, there will also be a more specific, quantitative guide to the type and scale of project which might require environmental assessment. For each of the types of development set out in the new Annex II there will be a published set of 'exclusive' thresholds and 'indicative' thresholds. The purpose of the exclusive thresholds is to focus resources on those projects which could have a significant effect on the environment; where a project falls below all the exclusive thresholds, there will be a clear presumption that environmental assessment is not required. (However, it is proposed that the exclusive thresholds will not apply for projects set in the most sensitive locations, e.g. international or nationally designated protection areas such as sites of special scientific interest (SSSIs).) Where a project is above any of the exclusive thresholds, a formal decision on the need for environmental assessment will be necessary, and the indicative thresholds will be relevant to the decision on whether the effects of the development will be significant enough to require environmental assessment. It is acknowledged that it will never be possible to formulate a precise threshold which can provide a simple, universally applicable test of whether a project will have significant effects. However, the indicative thresholds will be an aid to decision makers and developers alike. Initial proposals for the exclusive thresholds and indicative thresholds are set out in Appendix B of the Second Consultation Paper.

Scoping and the content of environmental statements

Under the new Article 5(2), developers have the power to ask the competent authority for formal advice on the scope of the information required in a particular environmental statement. The competent authority must then consult the developer and organisations with relevant environmental responsibilities before giving its opinion. Member states are also given the option to make it mandatory for competent authorities to give an opinion on the scope of every proposed environmental statement, whether requested by the developer or not.

The UK Government has decided not to introduce a system which makes 'scoping' of the environmental statement mandatory in every case, as this would limit flexibility and possibly create unnecessary delays due to excessive formality. Instead, the First Consultation Paper states that government guidance will continue to encourage informal consultation prior to the preparation of an environmental statement and, in addition, competent authorities will continue to have the power to request further information from a developer following the submission of an environmental statement. To comply with the 1997 Directive, legislation will be introduced to allow a developer to require a competent authority to give a formal opinion on the information that should be included in a particular environmental statement.

A further change in the content of environmental statements is that the new Article 5(3) makes it mandatory for environmental statements to include an outline of the main alternatives studied by the developer, and an indication of the main reasons for the developer's choice, taking into account the environmental effects. The EA Regulations will be amended to implement this.

Other changes

There is an added requirement that for every Annex II project the relevant competent authority must now make public its decision on whether environmental assessment is needed *or is not needed*. This requirement is added by the new Article 4(4). Under the present town and country planning system, such decisions are only made available to the public when the developer asks the local planning authority for an opinion in writing, or in any circumstance where the Secretary of State gives a direction on the matter. Potentially this new requirement would impose an additional burden on the various competent authorities if a decision had to be publicised for every project (or modification to a project) which fell within Annex II. However, the proposed 'exclusive thresholds' would substantially reduce this burden, as smaller projects would be excluded altogether from the need for consideration. The proposal in the First Consultation Paper is that, for all projects above the exclusive thresholds, the local planning authority will be required to place on the planning register a record of the decision, including that by the Secretary of State, on the need for environmental assessment.

The 1997 Directive also introduces the requirement for the competent authority to make public the main reasons for its decision on whether or not to grant development consent for any project which has been subject to an environmental assessment. This is a change from the present system, where reasons are normally given only where an application is refused.

The procedures for consultations with other member states where a project is likely to have significant transboundary environmental effects are enhanced by the 1997 Directive. This is provided for in the new Article 7 and brings the EA Regulations into line with the United Nations Economic Commission for Europe Convention on Environmental Impact Assessment in a Transboundary Context (the 'Espoo Convention').

There is also a new provision which gives member states the option to create a single procedure which could simultaneously meet the requirements of both the Amended EA Directive and Directive 96/61/EC on Integrated Pollution Prevention and Control (IPPC Directive). This is in recognition of the fact that a number of projects which require environmental assessment will also be covered by the new IPPC Directive; time and resources may be wasted if two separate applications are required and (as may happen) environmental organisations are consulted twice. In its Second Consultation Paper on the implementation of the IPPC Directive, the Government noted the need to ensure as much coherence as possible between the arrangements made to implement IPPC and the 1997 Directive, but it did not put forward any concrete proposals, preferring to wait for further consultation.

4

The environmental assessment process

Introduction
As explained in Chapter 3, the requirement to carry out an environmental assessment under UK law (referring to the law of England and Wales) arises principally in two ways:

- Under town and country planning legislation where a consent for development is required. In this case, the principal regulations of relevance are The Town and Country Planning (Assessment of Environmental Effects) Regulations 1988 (as amended) (the 'EA Regulations'). References in this chapter to 'Regulation(s)' are to the individual regulations in the EA Regulations.
- Under other legislation for projects which do not necessarily fall within the town and country planning system. A summary of projects which fall into this category and the applicable legislation is given in the last section of this chapter.

Environmental assessment and town and country planning

Introduction to the development consent process and town and country planning legislation
Most environmental assessments are carried out as part of the process for obtaining planning permission.

Planning legislation has created a complex system for the control of land use together with the design and form of the built environment. Accordingly, planning law has an essential part to play in the protection of the environment. The current legislation revolves around the Town and Country Planning Act 1990 (the 'Act'), which is supported by a great deal of secondary legislation as well as national and regional guidance produced by the Government and local policy produced by individual authorities.

The general rule, however, is that all 'development' of land requires the granting of planning permission by the competent authority. Development is defined in the Act (Section 55 (1)) as:

> the carrying out of building, engineering, mining or other opera-
> tions in, on, over or under land, or the making of any material
> change of use of any buildings or other land.

There has been much judicial interpretation of the meaning and application of this definition, the consideration of which is beyond the scope of this book. It is important to note, however, that there are a number of exceptions to and exemptions from this general rule. Certain types of operations are deemed by law not to be 'development' and so do not require planning permission. For example, the carrying-out within the boundaries of a road of any works required for the maintenance or improvement of the road is not development and nor is the carrying-out of works which affect only the interior of a building (Section 55(2) of the Act).

Another exemption from the general rule is that certain categories of development, set out in the Town and Country Planning (General Permitted Development) Order 1995 (SI 1995 No. 418 as amended) (known as the 'GPDO'), are automatically granted planning permission. Permission of this sort is known as 'deemed permission' and the rights to carry out development under deemed permissions are known as 'permitted development rights'. The general types of activity that are given permitted development rights under the GPDO are developments which are minor, which are carried out by a wide range of public services or which are favoured activities, especially agriculture and forestry projects. Projects which were subject to permitted development rights used to escape altogether the need for environmental assessment as the granting of planning permission was automatic. However, due to pressure from the European Community which considered this to be inadequate implementation of the EA Directive, the GPDO now withdraws permitted development rights wherever a development falls within Schedule 1 of EA Regulations, or Schedule 2 where the development is likely to have significant environmental effects. The result is that planning permission must be obtained in the usual way for developments with permitted development rights wherever environmental assessment is required.

Separate regulations have been brought into force which set down specific procedures to be followed where projects which would otherwise benefit from permitted development rights now fall within the environmental assessment regime (see The Town and Country Planning (Environmental Assessment and Permitted Development) Regulations 1995 (SI 1995 No. 417). Under these procedures the developer may apply to the local authority for an opinion as to whether the proposed project (which would otherwise benefit from permitted development rights) should be subject to environmental assessment and hence require specific planning permission. There is a right of appeal to the Secretary of State. In addition, many of the types of development which benefit from permitted development rights are subjected to more specific procedures in relation to the assessment of their environmental effects under separate environmental assessment regulations (e.g. afforestation and fish farms). These are dealt with at the end of this chapter.

Applications for planning permission are made to the local planning authority, which will usually be the district (or borough) council, but may in certain cases be another body specified by the Act. For example, in London the relevant London Borough Council will usually be the local planning authority; in some cases an Urban Development Corporation will be the local planning authority; for certain types of development the County Council, rather than the District Council, will be the local planning authority.

The EA Regulations

Most environmental assessments take place under and in accordance with the EA Regulations, which were introduced for the purpose of implementing the EA Directive by the deadline of 3 July 1988. The EA Regulations came into effect on 15 July 1988, shortly after the deadline. The Government's aim in implementing the EA Directive was to avoid placing any additional cost, administrative or other burdens on developers or local planning authorities. The Government did not intend to impose legislation which would require environmental assessment in cases where the EA Directive does not require assessment. Accordingly, the tenor of the EA Regulations is to achieve the minimum necessary to implement the EA Directive. Parts of the EA Regulations reproduce, almost exactly, parts of the EA Directive.

The EA Regulations should not be considered in isolation. Guidance on environmental assessment and the EA Regulations has been produced by the Department of the Environment. The two main limbs of such guidance are:

- guidance to local planning authorities contained in the DoE Circular 15/88 (the 'Circular'), now supplemented by DoE Circular 7/94; and
- guidance to developers contained in *Environmental Assessment — A Guide to the Procedures* (DoE and Welsh Office, 1989 — the 'Blue Book').

Although the Circular and its update, Circular 7/94, are aimed at providing guidance to local planning authorities, in practice it is also used by developers, their advisors and the public as an aid to understanding the meaning and requirements of the EA Regulations and the EA Directive.

The EA Regulations set out the requirements of and the procedures to be followed to determine whether an environmental assessment must be carried out before development can be authorised by planning permission granted under the Act, and to prepare and submit an environmental statement where one is required. The projects listed in Annexes I and II of the EA Directive (see Appendix 1) are reproduced almost verbatim in Schedules 1 and 2 of the EA Regulations (see Appendix 2). Nevertheless, Schedule 1 amplifies and changes the description of projects listed in Annex I in order to explain them in recognised terms. Due to these subtle differences, compliance with the EA Regulations will not automatically mean compliance with the EA Directive, since large parts of the wording of the EA Regulations do not derive directly from the EA Directive. However, it is at least arguable that development must comply with both the EA Regulations and the EA Directive, as explained in the section of Chapter 3 on the doctrine of direct effect. Therefore where the Regulations and Directive diverge, it is prudent to comply with whichever contains the more stringent requirement.

The EA Directive has recently been amended and amplified by the Directive 97/11/EC and the UK Government is currently engaged in consultations to prepare for the implementation of the changes brought about by this new Directive. The changes that have been made to the EA Directive and the proposals for implementing those changes into the UK regulations as set out in the government's two Consultation Papers were discussed in the last section of Chapter 3. There will clearly be some significant changes, including the introduction of exclusive thresholds and further indicative criteria. However, the rest of this chapter deals with the environmental assessment process under the UK regulations which are currently in force, as it is these provisions that are relevant to any application for development consent made until the time when the changes are implemented. The date by which member states are obliged to implement the changes is 14 March 1999, according to the 1997 Directive. It is therefore advisable to read this chapter in conjunction with the last section of Chapter 3 in order to understand how the environment assessment process in the UK will change in the future.

As with much environmental legislation, an understanding of the EA Regulations depends upon the definitions and terms as set out in Regulation 2(1). Paradoxically, the term 'environmental assessment' itself is neither defined by nor used in EA Regulations. Instead, the EA Regulations define and use the terms 'environmental information' and 'environmental statement' throughout. Environmental information consists of all the information gathered during the first stages of the environmental assessment process. Environmental information is defined (Regulation 2(1)) as:

> The environmental statement prepared by the applicant or appellant, any representations made by any body required by these Regulations to be invited to make representations or to be consulted and duly made by any other person about the likely environmental effects of the proposed development.

Therefore, environmental information includes information gathered and evaluated by the developer — which should be presented in a document (or series of documents) called the environmental statement — as well as the information derived from the consultation process that is required by the EA Regulations. The environmental statement must contain the information specified by the EA Regulations (Regulation 2(1) and Schedule 3, paragraph 2), including a summary in non-technical language. In this chapter, 'environmental information' and 'environmental statement' mean the same as in the EA Regulations. Any reference to environmental assessment refers to the whole process of consultation and assessment required under the EA Regulations or, in some cases, other legislation.

Environmental assessment — the basic requirement

The EA Regulations apply to applications for planning permission made on or after 15 July 1988. Where there is an application for planning permission for a development of the type which (under the EA Regulations) requires an environmental assessment, then the EA Regulations (Regulations 4(1) and 4(2)) prohibit the granting of planning permission in respect of such development unless the decision maker — be it the local planning authority, an inspector or the Secretary of State — has:

- taken the environmental information into consideration in making any decision with regard to the granting of planning permission; and
- stated in the decision that the environmental information has been taken into consideration.

It follows from this that the purported granting of planning permission without an environmental assessment where one is required under the EA Regulations is not valid. The first decision to make, therefore, is whether a particular project falls within the EA Regulations, and so requires an environmental assessment. This decision is for the planning authority in the first instance, but in most circumstances a prospective developer will wish to go through a similar analysis so as to anticipate the planning authority's decision and ensure that the call for an assessment is justified.

Projects to which the EA Regulations apply

The general rule
The general rule is that an environmental assessment is required in respect of any type of development listed in:

- Schedule 1 of the EA Regulations (see Appendix 2); or
- Schedule 2 of the EA Regulations *and which would be likely to have significant effects on the environment by virtue of its nature, size or location* (see Appendix 2).

The distinction between Schedule 1 and Schedule 2 projects is similar to the distinction created by the EA Directive between Annex I and Annex II projects (see Chapter 3). If a project is listed in Schedule 1 an environmental assessment *must* be carried out. If, however, a project is listed in Schedule 2 an environmental assessment will only be required where the project in question is likely to give rise to significant effects on the environment, by virtue of its nature, size or location.

For Schedule 1 projects, therefore, the position is usually clear. For Schedule 2 projects, however, the need for environmental assessment is not as easy to ascertain. An evaluation must be made at an early stage, without the benefit of having carried out the environmental assessment itself. One must consider whether or not the project is likely to have 'significant effects on the environment', even though neither 'significant effects' nor 'environment' is defined by the EA Regulations. There is, however, some help at hand in the Circular, which is discussed below.

In short, the need for environmental assessment turns not only upon the inherent characteristics of the development itself, but upon the sensitivity of the area involved.

The flow chart in Fig. 2 shows the decision process to be followed in deciding whether a given development will require environmental assessment.

Schedule 1 or Schedule 2 projects

Whether a project falls within the terms of Schedule 1 or Schedule 2 can depend upon various matters discussed below.

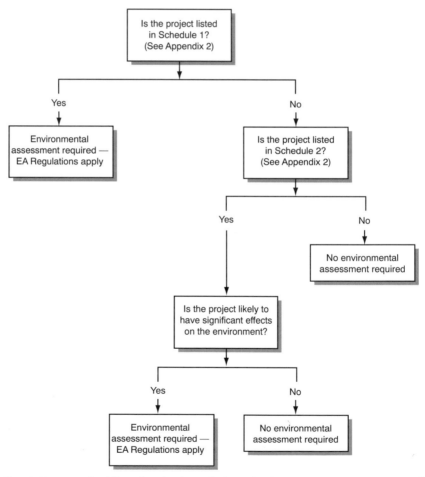

Fig. 2. Process to be followed in deciding whether a development requires environmental assessment

Directions of the Secretary of State
Under Regulation 3 and Section 74 of the Act the Secretary of State
has a general power to make directions, which may be included within
'development orders', upon the manner in which applications for planning
permission are treated. Thereby, the Secretary of State may prescribe
that a particular proposed development is:

- exempt from the provisions of the EA Regulations, even though it
  falls within Schedule 1 or Schedule 2; or
- exempt from the requirement for consideration of environmental
  assessment prior to the granting of planning permission — this
  provision effectively permits the Secretary of State to restore
  permitted development rights to a development so that it does not
  require environmental assessment; or
- subject to the requirement for consideration of environmental
  assessment.

Deemed Schedule 1 or Schedule 2 projects
In addition to the powers to make certain directions as discussed above,
there are three circumstances in which a project can be *deemed* to be a
Schedule 1 or Schedule 2 application:

- when the applicant and local planning authority agree that that is
  the case (Regulations 4(3) and 4(4)(iii));
- when the applicant for planning permission volunteers an environmental statement which is expressed to be for the purposes of the
  EA Regulations (Regulations 4(3) and 4(4)(i));
- when the applicant for planning permission fails to apply to the
  Secretary of State for a direction in a case where the local planning
  authority is of the opinion (see below) that consideration of
  environmental information is required (Regulations 4(3) and
  4(4)(ii)).

Seeking opinion from the local planning authority
Regulation 5 provides that anyone who is intending to apply for planning
permission can, before making such an application, apply to the local
planning authority requesting a written opinion on whether or not the
development falls within Schedule 1 or Schedule 2. If, in the opinion
of the local planning authority, the proposed development falls within
Schedule 2, then whether such development is likely to have significant
effects on the environment must be considered in order to decide whether
an environmental assessment is needed or not. The procedures for
requesting such an opinion are set out in greater detail below.

Schedule 2 development — is an environmental assessment needed?
Leaving aside circumstances where the general position has been altered
by a direction or a deemed direction, whether a Schedule 2 project must
be subject to environmental assessment depends upon the likelihood of
significant effects on the environment arising. This is the basic and
fundamental test to be carried out in each case, and guidance is offered
in the Circular and the Blue Book. In particular, the Circular makes
three non-exclusive suggestions about Schedule 2 projects for which
environmental assessment will be needed, to assist with this process of
assessment and preliminary decision making.

- '. . . *major projects which are of more than local importance*'
  Projects of more than local importance are projects which, merely
  by scale, are significant and likely to have far-reaching environmental
  effects. These projects will always justify an environmental assessment.
  Examples include large mining operations, substantial new manufacturing plant within the categories listed in Schedule 2, and major
  infrastructure projects (Circular, paragraphs 22–23).
- '. . . *projects on a smaller scale which are proposed for particularly
  sensitive or vulnerable locations*'
  The more environmentally sensitive the location, the more likely
  that environmental effects will be significant and hence the more
  likely that the project will require environmental assessment.
  Schedule 2 projects which are likely to have significant effect include
  those likely to affect the special character of areas protected by law
  by reason of their natural, cultural or historical significance. These
  areas are typically national parks, areas of outstanding natural beauty
  (AONBs), sites of special scientific interest (SSSIs), national nature
  reserves or monuments of major archaeological importance. There
  is, however, no automatic presumption that environmental assessment
  will be needed simply because a proposed development is located in
  a protected area. Rather, the test is whether the project is likely to
  have significant effects on that particular area's environment. Note
  also that consideration of location is not limited to the natural
  environment — projects in urban locations may justify an environmental assessment where the proposed development would be likely
  to have significant effects on a densely populated area (Circular,
  paragraphs 24–27).
- '. . . *projects with unusually complex and potentially adverse environmental
  effects, where expert and detailed analysis of those effects would be desirable
  and would be relevant to the issue of principle as to whether or not
  development should be permitted*'.

76

Within the scope of this definition generally fall industrial or manufacturing premises which will give rise to emissions which are particularly hazardous to humans, nature or both. In such cases environmental assessment is necessary.

The view expressed in the Circular is that a minority of projects will fall within any of these three categories and that in practice it will not be difficult to decide whether or not an environmental assessment is necessary. Even in borderline cases there is often little to be gained from resisting the performance of an environmental assessment. One is likely only to cause delays to the planning process, perhaps aggravating public opinion into the bargain. It may be prudent, therefore, for a developer to carry out an environmental assessment even if the strict requirements of the EA Regulations are in doubt.

In addition, indicative criteria and thresholds are set out in Appendix A of the Circular (and repeated in Appendix 2 of the Blue Book) which can be used to identify Schedule 2 applications. They are summarised in Table 8. If the criteria or thresholds are met or exceeded an environmental assessment will usually be called for, but these criteria and thresholds are not to be regarded as absolute. They are not set out in the EA Regulations, although their existence in the guidance is persuasive to decision makers. The rationale for using indicative rather than legally binding criteria and thresholds appears to be that it is not possible to produce reliable criteria and thresholds applicable to all circumstances which will always provide a simple and reliable test of need for an environmental assessment. Therefore, as the Department of the Environment (Circular, paragraph 30) have stated,

> The most that such criteria can offer is a broad indication of the type or scale of project which may be a candidate for assessment — and conversely, an indication of the sort of project for which assessment is not likely to be required.

Note also that significant local opposition or controversy is immaterial so far as the procedure is concerned.

Obtaining an opinion on whether an EA is needed

As discussed above, applicants for planning permission may apply to the appropriate planning authority for a written opinion on whether the proposed development falls within Schedule 1 or Schedule 2, pursuant to the procedure set out in the EA Regulations. The procedure is voluntary and should be carried out prior to application for planning permission.

Table 8. Summary of indicative criteria and thresholds for identifying Schedule 2 projects which require environmental assessment

Schedule 2 project	Indicative criterion/threshold
Agriculture[a]	
New pig-rearing installations	Generally no EA except for housing for more than: • 400 sows; or • 5000 fattening pigs.
New poultry-rearing installations	Generally no EA except for housing for more than: • 100 000 broilers; or • 50 000 layers, turkeys or other poultry.
Salmon farming	Depends upon the environmental effects generally and in particular on implications for the river system. Developments designed to produce less than 100 t of fish per annum should not normally require EA.
New drainage and flood defence works	EA may be required; consultations with drainage and environmental authorities indicate that the project is likely to have a significant environmental effect.
Extractive industry[a]	
Mineral workings	Need for EA will depend upon factors such as: • sensitivity of location; • size; • working methods; • proposals for waste disposal; • nature and extent of processing and ancillary operations; • arrangements for transporting minerals from site; • duration of the proposed workings.
Mineral applications in national parks or areas of outstanding national beauty	Generally will require EA.
Deep mines	May require EA.
Open-cast mines and sand and gravel workings	May require EA if more than 50 ha. May require EA if less than 50 ha if in a sensitive area or likely to involve obtrusive operations.

Table 8 continued.

Schedule 2 project	Indicative criterion/threshold
Rock quarries or clay operations	Depends on location, scale and type of activities.
Oil and gas extraction	Considerations will include: ● volume of oil/gas to be produced; ● arrangements for transport from site; ● sensitivity of area affected; ● EA may be necessary where 300 t or more per day are produced or if there is site sensitivity to disturbance from normal operations. Exploratory deep drilling is likely to require EA only if in a sensitive location or if the site is sensitive to limited disturbance occurring over a short period.

Manufacturing industry[a]	
New manufacturing plant	20–30 ha or above may require EA. In addition EA may be required for such plant because of discharge of waste or emission of pollutants. Factors to be taken into account include: ● whether the project includes an APC process; ● whether the project includes a process which involves discharges to water which require the consent of the water authority; ● whether the installation gives rise to environmentally significant quantities of hazardous or polluting substances; ● whether the process gives rise to radioactive or other hazardous waste.

Industrial estate development projects[a]	
Industrial estate development	May require EA where: ● the proposed site area is in excess of 20 ha; or ● there are more than 1000 dwellings within 200 m of the site boundary of the proposed estate. A smaller estate may require EA in sensitive, urban or rural areas.

Table 8 contiuned

Schedule 2 project	Indicative criterion/threshold
Urban development projects[a]	
Redevelopment of previously developed land	Not likely to require EA unless: • the proposed use is a development listed in Schedule 1 or Schedule 2; or • the project is on a much greater scale than the previous use of the land.
New urban development schemes on sites which have not previously been intensively developed	Depends on sensitivity of locations. May require EA where: • the site area is more than 5 ha in an urbanised area; or • there are more than 700 dwellings within 200 m of the site boundary of the proposed development; or • the development would provide a total of more than 10 000 m^2 (gross) of shops, offices or other commercial uses. High-rise development above 50 m may require consideration for EA.
Small urban development schemes	May require EA in particularly sensitive areas, e.g. central-area redevelopment schemes in historic town centres.
Major out-of town shopping schemes	The need for EA should be considered in the light of sensitivity of the particular location; a floor area threshold of about 20 000 m^2 may provide an indication of significance.
Local roads[a]	
Construction of new roads and major road improvements outside urban areas	EA may be required where such development is for a road: • over 10 km long; or • over 1 km long, if it passes through a National Park; or • through or within 100 m of an SSSI, a national nature reserve or a conservation area.

Table 8 continued

Schedule 2 project	Indicative criterion/threshold
Within urban areas	EA may be required where there are more than 1500 dwellings within 100 m of the centre line of the proposed road (or an existing road in the case of major improvements).
Airports[a]	
Construction of airport	EA required under Schedule 1 if the airport has a basic runway length of over 2100 m.
Smaller new airports	Will generally require EA.
Major works, e.g. new runways or passenger terminals at larger airports	May require EA if the original development would have required EA under Schedule 1.
Other infrastructure projects[a]	Projects requiring sites in excess of 100 ha may require EA.
Waste disposal[a]	
Installations for treatment or disposal of household, industrial and commercial waste, including landfill sites	May require EA for a capacity of more than 75 000 t a year. Sites taking smaller tonnages and sites which will only accept inert wastes are unlikely to require EA unless they are in sensitive locations.
Wind generators[b]	May well require EA if: • the development is located within or is likely to have significant environmental effects on a National Park, the Broads or the New Forest, an AONB, an SSSI or heritage coast; or • the development consists of more than ten wind generators; or • the total installed capacity of the development exceeds 5 MW.
Motorway service areas[b]	May well require EA where the proposed location is in a National Park, the Broads or the New Forest, an AONB or an SSSI; and for any such development over 5 ha in areas outside those locations.

Table 8 continued

Schedule 2 project	Indicative criterion/threshold
Coast protection works [b]	May well require EA where the coast protection works are proposed to be located in or are likely to have significant effects on a National Park, the Broads or the New Forest, an AONB, an SSSI, heritage coast or a marine nature reserve.

[a] DoE Circular 15/88. [b] DoE Circular 7/94.

Planning authorities should give their written opinion upon:

- whether or not a development falls within Schedule 1 or Schedule 2; and
- if in the opinion of the local planning authority a proposed development falls within Schedule 2, whether it is likely to have significant effects on the environment so as to require an environmental assessment.

The applicant for an opinion has a wide discretion over the amount and type of information which is provided to the planning authority for this purpose, but the following information must be included:

- a plan sufficient to identify the site of the proposed development; and
- a brief description of the nature and purpose of the proposed development and its possible effects upon the environment.

Otherwise, applicants need provide only such other information or representations as they wish, although the planning authority may notify them of particular information which it requires before being able to give an opinion.

In the Circular it is envisaged that no more than information sufficient to consider the broad significance of issues raised by the proposal will be required, and an authority should not require the same information as would be necessary for consideration of a planning application. When the authority has sufficient information to consider the application for an opinion it should follow the reasoning process as set out in Fig. 2 to determine whether the development is one which requires environmental assessment.

The EA Regulations do not oblige the planning authority to consult any private or public body or individual before giving its opinion. Whether a development is a Schedule 1 or Schedule 2 development is usually fairly clear. However, where a local planning authority is uncertain as to whether or not a Schedule 2 project is likely to have significant effects on the environment, it would be well advised to seek advice from relevant expert bodies. In borderline cases an authority may request guidance from such bodies as the Environment Agency or the Royal Society for the Protection of Birds (RSPB).

The authority should have formulated its opinion and notified the developer of it within three weeks of receiving the request from the developer. The guidance in the Circular does point out, however, that the authority and the developer may extend this period by agreement. If the planning authority concludes that consideration of environmental information is required before planning permission can be granted, it must provide a written statement setting out full reasons for this conclusion, with its opinion (Regulation 5(4)). A copy of any written opinion must be placed on the public register and be available for public inspection.

To make a request for an opinion requires a certain amount of preparation and expenditure at the outset by the developer, who may thus be deterred from using the procedure. On the other hand, the procedure offers several distinct advantages:

- An early indication to the developer and to the local planning authority of the significance of the environmental issues raised by the development proposal may render the proposal more acceptable to the local planning authority and the public in the long term, bearing in mind that public involvement in a proposed development at an early stage can be influential.
- The request for an opinion in advance of an application for planning permission makes the local planning authority well aware of a proposed development at an early stage in the development process.
- Any opinion received from the local planning authority allows a developer to amend a draft planning application to make it more acceptable to that authority. This is particularly useful if significant environmental issues are identified or confirmed by the opinion given by the authority.
- The information provided with the request for an opinion will help to identify the scope of the environmental information that will need to be gathered for the environmental statement. This will give an indication of any potentially significant environmental impact

and in turn allow the design of the development in question to be developed or adapted to minimise that impact.

- The written opinion of the local planning authority will give the developer an initial ruling as to whether or not an environmental assessment is required. If the opinion is that environmental assessment is not required, it will show clearly that environmental issues were considered but not felt to be significant.

Challenging the planning authority's opinion — applying for a direction

In two cases where a developer is not satisfied with the position of the planning authority, an application can be made to the Secretary of State for a direction upon the matter. These are:

- where an authority gives an opinion that the consideration of environmental information is required before planning permission can be granted, but the developer disagrees (Regulation 5(6)(a)); and
- where an authority fails to give an opinion within three weeks (or such longer period as may have been agreed), even though the applicant has not failed to supply any information requested by the authority (Regulation 5(6)(b)).

An application for a direction must be accompanied by a copy of the original request made to the planning authority and accompanying documents; a copy of any request by the authority for further information with any response made; any opinion given by the authority and statement of reasons for that opinion; and any representations which the developer may wish to make (Regulation 6(1)). Copies of any application and representations made by a developer to the Secretary of State should be sent to the local planning authority (Regulation 6(2)).

The Secretary of State, when in receipt of the application, may feel that he needs more information and he is entitled to request it from the applicant. He may also request that the local planning authority provide information on any particular points he wishes to raise. The Secretary of State should give his direction within three weeks of the application unless it is reasonable for him to require further time to do so. Where the Secretary of State's direction is to the effect that the proposed application would be a Schedule 1 or Schedule 2 application, he must give full, clear and precise reasons for his conclusion.

The Secretary of State's direction given under Regulation 6 overrides the local planning authority's opinion and is binding. If no direction is applied for, then the local planning authority's opinion is binding. Opinions may only be challenged in the High Court in circumstances where procedures have not been followed, or the direction made or opinion given is manifestly unreasonable.

Having received an opinion or direction, the developer will know, before preparing and submitting the planning application, whether the development in question requires the preparation of an environmental statement at that point. However, this is not always the end of the story. Before an application for planning permission is actually decided, environmental assessment may have become necessary. It is conceivable, therefore, that a local planning authority might have given a perfectly proper opinion that no environmental statement was required, but during the preparation or consideration of the planning application it transpires that the proposed development is in fact one requiring environmental assessment. In cases such as these, the general rule in Regulation 4(2) applies, namely that the planning permission cannot be granted without consideration of the environmental information.

Facilitating preparation of environmental statements

Where an environmental statement is to be submitted, even if on a voluntary basis, Regulation 8 of the EA Regulations sets out a formula whereby information must be made available to the developer by the authority and certain other public bodies.

Part of the procedure set out in Regulation 8 for the facilitation of the preparation of environmental statements includes the process of notification by the local planning authority to various prescribed bodies or organisations, who are thereby placed under a duty to provide information to the prospective applicant for planning permission. The local planning authority has this duty to give notice to other interested bodies in several circumstances where the need for an environmental statement has arisen:

- where written notice is given of the intention by a prospective applicant for planning permission to make a Schedule 1 or Schedule 2 application and to submit an environmental statement with that application; or
- where the applicant for planning permission has agreed or conceded in writing that the submission of an environmental statement is required pursuant to Regulation 4(4)(iii) (see above); or
- the Secretary of State has directed that a proposed application for

planning permission would be a Schedule 1 or Schedule 2 application pursuant to Regulation 6.

The public bodies obliged to make relevant information available to a developer are prescribed by Regulation 8, read together with Article 10 of the Town and Country Planning (General Development Procedure) Order 1995 (SI 1995 No. 419) (known as the 'GDPO'). The local planning authority should make known to them the name and address of the prospective applicant, while informing the prospective applicant of the bodies that have been notified. In addition to the bodies that under normal planning procedures must be consulted pursuant to Article 10 of the GDPO, the other bodies to be notified by the authority are:

- any principal authority for the area where the land is situated, if not the local planning authority, for example the County Council;
- the Countryside Commission;
- the Nature Conservancy Council (now divided into the Nature Conservancy Council for England (known as 'English Nature'), the Nature Conservancy Council for Scotland and the Countryside Council for Wales);
- the Environment Agency in respect of proposed development which in the opinion of the local planning authority will involve:
  - mining operations;
  - the manufacturing industry; or
  - the disposal of waste.

and is likely either to give rise to radioactive waste, the disposal of which requires authorisation under the Radioactive Substances Act 1993, or discharges which are controlled as special waste or are likely to require the licence for consent of a sewerage undertaker in respect of a discharge; or to involve works specified in Schedule 1 to the Health and Safety (Emissions to the Atmosphere) Regulations 1983.

Procedures for requiring the submission of an environmental statement

Where the local planning authority receives a planning application which it believes may possibly be a Schedule 1 or Schedule 2 application and which is not accompanied by an environmental statement, it has the power to issue a notice requiring that an environmental statement is submitted (Regulation 9(1)). Clearly, before issuing such a notice the local planning authority should check its records to establish whether it has given a written opinion on the need for an environmental statement, and whether the Secretary of State has issued any direction.

The issuing of a notice under Regulation 9 by the local planning authority must be made within three weeks of receipt of the planning application or such longer period as may be agreed. It must set out in full the reasons why it has been issued and allow three weeks for a response. Broadly speaking, the applicant may respond in either of two ways:

- by stating that it accepts that an environmental statement is needed and that a statement will be provided; or
- by stating that the applicant intends to write to the Secretary of State requesting a direction on the matter.

The applicant for planning permission should not, however, ignore the notice. A failure to respond within the three-week period results in the permission sought being automatically deemed to have been refused at the end of the three-week period. Furthermore, if planning permission is deemed to be refused, the applicant is then prohibited from making an appeal against the deemed refusal under Section 78 of the Act (Regulation 9(3)).

In some cases the Secretary of State, rather than the local planning authority, deals with and decides the planning application — for instance, when the Secretary of State exercises the power to 'call in' an application under Section 77 of the Act for determination himself. In this case, under Regulation 10 the Secretary of State has similar powers and duties to those of the local planning authority. He may serve a notice requiring the submission of an environmental statement, the recipient of which then has three weeks within which to inform the Secretary of State in writing that an environmental statement will be provided. If the applicant fails to do so, then the Secretary of State has no further duty to deal with the application and must write to inform the applicant that no further action is being taken on the application (Regulation 10(3)).

Regulation 11 provides that the powers and duties of the Secretary of State under Regulation 10 outlined above similarly apply (subject to any necessary modifications) where he has received an appeal against a planning decision under Section 78 of the Act. If he considers that the proposed development falls within Schedule 1 or Schedule 2 so that consideration of environmental information is required before the planning appeal could be allowed, he can require the submission of an environmental statement. Again, if the applicant does not agree to supply an environment statement within three weeks, the Secretary of State has no further duty to deal with the application.

If the question of whether an application falls within Schedule 1 or Schedule 2 arises in the course of a planning appeal under sections 78 or 79 of the Act being dealt with by an inspector, the procedure is very similar. The inspector must refer the matter to the Secretary of State for a direction and until this is received the inspector may not allow the appeal, but can dismiss it. If the Secretary of State directs that it is a Schedule 1 or Schedule 2 application he will send a reasoned statement in writing which should explain his conclusions clearly and precisely. As before, the applicant has three weeks in which to agree to provide an environmental statement. Without such agreement the Secretary of State must notify the applicant, at the end of the three-week period, that no further action is being taken upon the appeal (Regulations 11(2)–(6)).

If, in any of the situations where the Secretary of State is responsible for a decision upon a planning application, the development is found to fall within Schedule 1 or Schedule 2 and an applicant does not provide an environmental assessment, the application can be disposed of in only one way. The EA Regulations dictate that unless the applicant submits an environmental statement and complies with Regulation 13(5), the Secretary of State or inspector (as appropriate) may only refuse planning permission.

The content of an environmental statement

The EA Regulations do not make any stipulation about the exact form of an environmental statement, which may consist of more than one document. However, there is a list of information specified in paragraph 2 of Schedule 3 which must be included. The full text of Schedule 3 to the EA Regulations is set out in Appendix 3 of this book. It can be seen that the information specified reflects Annex III to the EA Directive. In particular, a description of the likely effects on the environment of the development must be included, explained by reference to the possible impact of the development on:

- human beings, flora and fauna;
- soil, water and air;
- climate;
- the landscape;
- the interaction between any of the foregoing;
- material assets; and
- cultural heritage.

Paragraph 3 of Schedule 3 gives a list of further information which may be included by way of explanation or amplification of any specified information, but which is not obligatory. Schedule 3 also requires that an environmental statement must include a summary in non-technical language of the information specified, so as to enable non-experts to understand its findings.

The Blue Book gives a checklist of matters to be considered for inclusion in an environmental statement; this is not legally binding but it is intended as a guide to the subjects that need to be considered. It stresses the distinction between, on the one hand, the construction and commissioning phases of a development, and on the other hand the operational phase, and specifies that the environmental effects arising from each should be considered separately. Also, where the operational life of a development is expected to be limited, the effects of decommissioning or reinstating the land should be considered separately. The guidance in this checklist is presented under the following headings:

- information describing the project;
- information describing the site and its environment;
- assessment of effects;
- mitigating measures;
- risk of accidents and hazardous development.

The guidance given in the Blue Book states that the comprehensive nature of the checklist should not be taken to imply that all environmental statements should cover every conceivable aspect of a project's potential environmental effect at the same level of detail. While every environmental statement should provide a full factual description of the project, the focus should be on the main or significant effects to which a project is likely to give rise.

It should also be noted that, pursuant to Regulation 21, the authority to whom an environmental statement has been submitted may require additional information to be provided (see below).

It must be stressed that the legislation itself sets out only the formal requirements for the content of environmental statements, and that developers and authorities should discuss the scope of an environmental statement before its preparation is begun. While at present there are no legislative provisions relating to the 'scoping' of an environmental statement, the government intends to introduce provisions to encourage informal consultation in scoping when it brings in legislation to implement the changes to the EA Directive brought about by the 1997 Directive (as discussed in Chapter 3).

The content of environmental statements in practice is discussed in greater detail in the Chapter 5.

Publicity requirements

Publicity when a planning application is accompanied by an environmental statement

Where an application for planning permission is accompanied by an environmental statement, the EA Regulations do not provide specifically for the manner in which the application is to be publicised. Rather, the ordinary planning application procedures set out in the GDPO apply, and the onus to publicise a planning application falls upon the local planning authority.

The GDPO sets down a specific publicity procedure for Schedule 1 and Schedule 2 applications which are accompanied by an environmental statement (Regulations 8(2)(a) and 8(3), GDPO). The application must be publicised by:

- the posting of a notice on or near to the land to which the application relates, firmly affixed to some object so as to be easily visible and legible to the public for at least 21 days;
- serving a notice on any adjoining owner or occupier; and
- publication of the notice in a newspaper which circulates in the locality of the application site.

The form of the notice to be given is specified in Schedule 3 of the GDPO.

Publicity when an environmental statement is received in the course of a planning application

An environmental statement may be submitted after the planning application. This would be the case where:

- an applicant voluntarily submits an environmental statement expressed to be for the purpose of the EA Regulations;
- an applicant agrees or concedes that an environmental statement is required (Regulations 4(4)(ii), (iii));
- an applicant agrees to provide an environmental statement after

submission of the application and after notification by the local planning authority that in its opinion an environmental statement is required (Regulation 9(2)(i)); or

- the Secretary of State directs that an environmental statement is required (Regulations 10(2), 11(1), 11(5)).

In any of the above circumstances, special publicity requirements apply before the statement is submitted to the local planning authority (Regulation 13). The applicant is obliged to produce a notice which must be:

- published in the local newspaper circulating in the locality in which the land that is the subject of the planning application is situated (Regulation 13(2)); and
- posted on the land which is the subject of the planning application (unless the applicant does not have, and it is not reasonably practicable for him to acquire, rights enabling him to do so).

The notice should contain:

- the name and address of the applicant;
- the name and address of the local planning authority;
- the date on which the application was made (and if appropriate the date on which the application was referred to the Secretary of State for determination or appeal to him);
- the address or location and nature of the proposed development;
- a statement that a copy of the planning application, together with the plans and other documents submitted with it, including the environmental statement, may be inspected by members of the public at all reasonable hours. This should include:
  - o the address at which those documents can be inspected and the latest date on which they will be available for inspection — this must be not less than 20 days later than the date on which the notice was published, or first posted on the land, as appropriate;
  - o an address from which copies of the environmental statement may be obtained and a statement that copies may be obtained so long as stocks last and announcing whether any charge is to be made for copies and if so its amount;
  - o a statement that any person wishing to make representations about the application should do so in writing before a specified date.

The environmental statement itself must be accompanied by a copy of the notice published in the local newspaper, this copy must be certified as having been published in a named newspaper on a specified date and as having been posted on the land in compliance with the above require-ments.

Where the applicant for planning permission has given notice of his intention to submit an environmental statement, the person determining that application must suspend consideration of the application or appeal until the environmental statement and the certificates relating to the notices have been received. The local planning authority, inspector or the Secretary of State may not take any further steps except to refuse the application until the receipt of that information. A further 21-day period following the date of receipt of the environmental statement and other documentation must elapse before the application can be determined.

Regulation 14 provides that when an environmental statement is submitted to the local planning authority, three additional copies of the statement must be provided, making a total of four. If a copy of the environmental statement (or any part of it) is to be served on any other body, for example a statutory consultee, the copy of the environmental statement must be accompanied by a copy of the planning application, including any relevant plan. The applicant must also inform the consultee that representations may be made to the local planning authority. The local planning authority should be informed of the name of every person who has been served with a copy of the environmental statement (Regula-tion 14(1)). If part only of an application is served upon any body other than the local planning authority, the authority should be informed of the part of the statement so served.

Procedure when a local planning authority receives an environmental statement

On receipt of an environmental statement the local planning authority must do a number of things. It must:

- ensure that a copy of the environmental statement is placed on the public register, together with the planning application (Regulation 14(2)(a));
- send three copies of the environmental statement and a copy of the planning application and any documents submitted with it to the Secretary of State (Regulation 14(2)(b)); and
- inform consultees who have not been furnished by the applicant with a copy of the environmental statement that this statement is to be taken into account in determining the planning application.

The local planning authority should enquire whether the consultees wish to receive a copy of the environmental statement and inform them that they may make representations. If any organisation does wish to receive a copy of the environmental statement the local planning authority must enquire whether the applicant will serve the required copies on each of those bodies himself, or via the authority. If the latter option is chosen, the authority will serve the environmental statements as appropriate (Regulations 14(2)(c,d,e))

The EA Regulations prohibit the local planning authority from determining a planning application until 14 days have elapsed from the last date on which a copy of the environmental statement was served.

Procedure when the Secretary of State receives an environmental statement

Where a Schedule 1 or Schedule 2 application is before the Secretary of State by reason of an appeal to him under Section 78 or 79 of the Act, or because the application has been referred to him under Section 77 of the Act, the procedure differs slightly.

- Four copies of the environmental statement should be sent to the Secretary of State. He will send one copy to the local planning authority, which must ensure that it is placed upon the register.
- The Secretary of State should be informed by the applicant of the persons upon whom copies of the environmental statement have been served and the part of the statement which was so served.
- Any other obligations of the local planning authority referred to in the last section above fall upon the Secretary of State.
- The Secretary of State must also advise any body stipulated in the Regulation 10 GDPO 1995 of the environmental statement, as the local planning authority would have notified consultees under Regulation 8(5).

Miscellaneous obligations and powers

The EA Regulations provide for miscellaneous additional obligations and powers of the applicant for planning permission.

- The applicant must keep a reasonable number of copies of the environmental statement at the address set out in the notices published or posted pursuant to the GDPO or Regulation 13 from which such copies can be obtained (Regulation 18).

- Where, pursuant to an application for planning permission which includes an environmental statement, the application is referred to the Secretary of State or becomes the subject of an appeal under the Act, the applicant must supply the Secretary of State with three copies of the environmental statement unless this has already been done (Regulation 19).
- The applicant may impose a reasonable charge for a copy of an environmental statement made available to a member of the public (Regulation 20(1)).

Further information and evidence

Once an environmental statement has been submitted to either the local planning authority, the Secretary of State or an inspector, that person has the power to require that the applicant or appellant, as the case may be, provides additional information (Regulation 21). The information specified may concern any matter which is required to be, or which may be, dealt with in the environmental statement. The requirement to provide further information only applies where the person making that request is of the opinion that the applicant or appellant could provide further information about any of the matters specified in paragraph 3 of Schedule 3 of the EA Regulations. This information should be what is reasonably required to give proper consideration to the likely environmental effects of the proposed development.

Time limits

Ordinarily, a local planning authority should make a decision upon a planning application within eight weeks of receipt of the application (Section 7(2) of the Act, and Regulation 20(2) GDPO 1995). Under the EA Regulations, however, this period is extended to 16 weeks in the case of an application where the submission of an environmental statement is required (Regulation 16(2)). Furthermore, in computing the time which has expired from receipt of a planning application to the issue of a decision, no account is to be taken of any time before a direction by the Secretary of State if the authority issued an opinion that an environmental statement was required but the application was submitted without one (Regulation 16(1)). It follows from this, of course, that where the 16-week period applies an appeal on the basis of non-determination cannot be made until that period has passed.

Environmental assessment where planning permission is not required

The law relating to environmental assessment is incorporated primarily into the planning law system. Certain projects, however, are not subject to the planning system. This could be because:

- they are projects which fall within the town and country planning system but which do not require planning permission since they benefit from permitted development rights under the GPDO. For example, in respect of many large infrastructure projects such as railway lines, trunk roads, ports and harbours, the authority to grant the development consent rests with the relevant Secretary of State rather than a local planning authority.
- they are projects which simply do not fall within the town and country planning system and in respect of which approval for development falls under different legislation (i.e. not the Town and Country Planning Act 1990). For example, harbour works carried out below the medium low-water mark fall outside local planning authority jurisdiction and therefore outside the Town and Country Planning Act 1990.

To deal with this, the Government has extended the provisions of the EA Directive to other projects by means of separate regulations. The following section sets out brief details of the regulations applying in England and Wales and the way in which these other projects are brought within the environmental assessment procedures.

It was noted at the beginning of this chapter that under the GPDO permitted development rights are now withdrawn for projects which would otherwise require environmental assessment under the EA Regulations. However, before 1995 this withdrawal did not exist and so separate regulations were enacted to introduce procedures for environmental assessment to projects with permitted development rights which were within the ambit of the EA Directive. The withdrawal of permitted development rights under the GPDO does not apply to those types of projects already subject to separate regulations.

The Environmental Assessment (Afforestation) Regulations 1988 No. 1207

These Regulations apply to afforestation projects and relate to decisions taken by the Forestry Commission. An afforestation project is a proposal for the initial planting of land with trees for forestry purposes.

The Regulations require that the Forestry Commission must not make a grant or loan for an afforestation project where it considers that the project will be likely to have significant effects on the environment and may lead to adverse ecological charges, without first considering environmental information.

An applicant may request an opinion upon whether environmental information is required by sending a map, a brief description and any further information of relevance to the Forestry Commission. The Commissioners may then request further information. The Commission has to give an opinion within four weeks of the request (or longer if agreed); if environmental information is required, reasons for requiring it must be given. On receipt of an opinion that environmental information is required, an applicant has four weeks to accept and propose an environmental statement or, if aggrieved, to apply to the Minister of Agriculture, Fisheries and Food for a direction. If no opinion is given, there is a presumption that environmental information is not required. To make an application to the Minister for a direction the applicant must enclose certain documents which are set out in Regulation 6. He may be required to provide further information. The Minister should respond within four weeks or such period as he considers reasonable, sending copies of his decision to the Commission and the applicant. In addition, the Minister may give a direction overturning a decision of the Commission in any case.

It will be noted that these procedures are very similar to those applying under the EA Regulations. Nevertheless, certain differences do exist. Prospective applicants to the Forestry Commission would be well advised to check the precise terms of the Regulations and any guidance provided by the Forestry Commission.

The Electricity and Pipeline Works (Assessment of Environmental Effects) Regulations 1990 No. 442

These Regulations (as amended by SI 1996/422 and SI 1997/629) re-enact the Electricity and Pipeline Works (Assessment of Environmental Effects) Regulations 1989 with certain modifications. They implement the EA Directive in relation to consents to construct, extend or operate a generating station and to install or keep installed an electricity line above ground in England and Wales, and authorisations to construct or divert oil or gas pipelines on land in Great Britain.

Once again, these Regulations are very similar in form, but not identical, to the EA Regulations. They state that the relevant Secretary of State shall not grant any of these consents or authorisations without prior consideration of environmental information. Except in relation to consents for nuclear generating stations or non-nuclear generating stations of heat output greater than or equal to 300 MW environmental statements are only required where the Secretary of State determines that the project is likely to have significant effects on the environment. A prospective applicant may seek the opinion of the Secretary of State on whether he considers an environmental statement to be necessary. There are also provisions similar to the EA Regulations relating to notice, consultation and procedure.

The Harbour Works (Assessment of Environmental Effects) Regulations 1988 No. 1336

These Regulations implement the EA Directive in relation to Harbour Revision Orders and Harbour Empowerment Orders concerning harbour works in England and Wales.

In short, they require the relevant Minister to assess whether any application for a harbour revision order falls within Annex I or II to the EA Directive.

If the Secretary of State determines that an environmental assessment is necessary, the applicant is required to apply with such relevant information as can reasonably be gathered. At a minimum, a description of the project proposed, measures to avoid or remedy significant environmental effects, data necessary to assess the main effects and a non-technical summary of all the information must be submitted.

The Harbour Works (Assessment of Environmental Effects) (No. 2) Regulations 1989 No. 424

These Regulations (as amended by SI 1996/1946) implement the EA Directive for England and Wales, and Scotland, in respect of harbour works below the medium low-water mark. The regime for these works falls outside the apparatus of the Town and Country Planning Acts because the works would be outside local planning authority jurisdiction, and therefore requires its own set of regulations.

Where an application or notice referred to in Regulation 4(1) is received, the appropriate Minister must reach a decision as to whether the Regulations apply as soon as reasonably practicable. Where they apply, the works may not be carried out unless the consent of the appropriate Minister is obtained or an environmental assessment is not required. In order to determine this, the Minister must assess whether the proposed works fall within Annex I or Annex II to the EA Directive and in the latter case whether their characteristics require that they should be subject to environmental assessment. To make this decision the Minister may require a brief description of the works; plans and sections showing the location, lines, situation and levels of the works; and such further information as he deems necessary.

If the Minister decides that the project does not require environmental assessment he must notify the developer and sometimes the Harbour Authority, and take no further action upon the application. In cases in which the proposed harbour works are to be subject to environmental assessment, the developer may be required to supply information referred to in Annex III of the Directive, which the Minister considers relevant and which the developer may reasonably be expected to supply. This information should always include a description of the works; of measures to avoid, reduce or if possible remedy significant adverse effects; data necessary to identify and assess the main environmental effects; and a non-technical summary of all the information.

The Environmental Assessment (Salmon Farming in Marine Waters) Regulations 1988 No. 1218

Some developments relating to on-shore salmon farming facilities may require planning permission. In that case, the EA Regulations will apply as for any other planning application. However, permission for salmon farming in marine waters, which are generally beyond the territorial jurisdiction of local planning authorities, is obtained by means of consents granted by the Crown Estate.

The Crown Estate Commissioners may not grant consents for salmon farming in marine water if, in their opinion, the development is likely to have significant effects upon the environment, unless they have considered environmental information in respect of the proposal.

The Highways (Assessment of Environmental Effects) Regulations 1988 No. 1241 and the Highways (Assessment of Environmental Effects) Regulations 1994 No. 1002

These Regulations implement the EA Directive for England and Wales in respect of proposals to construct or improve highways where the proposals are those of the Secretary of State. Regulations 1988/1241 insert Section 105A into the Highways Act 1980 and Regulations 1994/1002 supplement and clarify this by making amendments to Section 105A.

The Secretary of State must judge whether the project falls within Annex I or Annex II of the EA Directive and is such that it should be subject to an environmental assessment. If he determines that a proposal falls within these categories he must publish an environmental statement but not later than the publication of details of the project. The environmental statement must identify, describe and assess, in an appropriate manner, the direct and indirect effects of the project on those factors set out in Article 3 of the EA Directive and must do so in the light of each individual case. It must contain the information referred to in Annex III to the Directive where the Secretary of State considers it relevant and where the information may reasonably be gathered. The environmental statement must contain a description of the project and also a description of measures to avoid or reduce and, if possible, remedy any significant adverse effects.

Where an environmental statement is published the Secretary of State must give the public concerned an opportunity to express an opinion prior to initiation of the project. Similarly, statutory environmental bodies that are mentioned must be consulted prior to initiation of the project.

After considering the environmental statement and all opinions expressed, the Secretary of State must publish his decision on whether or not to initiate the project which must also state that he has considered the environmental statement and opinions.

The Transport and Works (Assessment of Environmental Effects) Regulations 1995 No. 1541

These Regulations amend Section 14 of the Transport and Works Act 1992 to implement Articles 8 and 9 of the EA Directive in relation to the construction or operation of railways, tramways, other guided transport systems, inland waterways and works interfering with rights of navigation.

Section 14 is supplemented to require the Secretary of State to confirm that prior to authorising or refusing a project he has considered any environmental statement and any opinions expressed on it.

The amendments also require the Secretary of State to send copies of the notice of determination for the project to all who made representations or raised objections where the Act does not already require this.

Further provision concerning submission and contents of environmental statements is given in the Transport and Works (Applications and Objections Procedure) Rules 1992 No. 2902.

The Land Drainage Improvement Works (Assessment of Environmental Effects) Regulations 1988 No. 1217
These Regulations (as amended by SI 1995/2195) implement the EA Directive for England and Wales in respect of land drainage improvement works proposed by the Environment Agency, or a water authority, internal drainage board or local authority ('drainage body'). New land drainage, flood defence and sea defence works require planning permission, and fall within the terms of the EA Regulations as a result. However, drainage bodies benefit from permitted development rights and consequently (before 1995) the EA Directive had to be specifically applied to improvement works proposed by them. These regulations supply a procedure for the provision of environmental statements relating to such improvements by drainage bodies where the body considers that by reason of, *inter alia*, their nature, size or location the proposed works are likely to have a significant effect on the environment.

5

Environmental assessment in practice

Introduction

Environmental impact assessment is essentially a tool to assist in the planning process. There is a legal requirement for the completion of environmental assessments for a variety of different types of project (see Chapter 3) that are considered as having the potential for causing significant environmental effects, but the undertaking of an environmental assessment will also enable the prediction of impacts (and their quantification) which will have the benefit of increasing the information held on a particular project. As a planning tool the environmental assessment process serves to inform interested parties (including the statutory consultees) of the likely environmental impacts of a proposed development and the methodologies that are to be used to mitigate or reduce the scale and significance of those impacts. There have been many definitions of environmental statements (e.g. Munn, 1979; Clark, 1987); the Government (Department of the Environment, 1994a) identifies the aim of an environmental statement (ES) as:

> to provide a full and systematic account of a development's likely effects on the environment, including those which are subject to pollution controls and the measures envisaged to avoid, reduce or remedy significant adverse effects.

The purpose of environmental assessment can therefore be defined as to serve as a management tool not only to assess impacts but to improve the quality of decisions (Formby, 1990; Wiesner, 1995). As a consequence, environmental assessment should be perceived not just as an interdisciplinary methodology that has to be followed to satisfy the regulations but also as an iterative process whereby the best and most appropriate decisions are made within the constraints of available resources.

The number of environmental statements submitted to planning authorities in support of planning applications has risen year by year since the introduction of the 1988 Regulations (Written Answers, May 1992). The estimate of the number of (ES) environmental statement documents likely to be submitted per year in the UK has risen from 190 in 1988 to over 320 in 1992 (University of Manchester, 1995a), and research undertaken at the Oxford Brookes University indicated that the actual number submitted in 1995 was 200 (Glasson, 1997). This rise is likely to continue with the probable inclusion of additional types of development for which an EA will be required (there have been six amendments to the original regulations since 1988).

There are several key stages and activities involved in the environmental assessment process. These are reviewed and discussed below.

Scoping

The purpose

Scoping of an environmental assessment is arguably the most important part of the process, next to the completion of the statement itself. Scoping as an activity has evolved in an attempt to rationalise the coverage of the EA and it is at the scoping stage that the issues that require study will be identified and the methodologies for undertaking the assessment of those issues agreed. The scoping study can confirm that an environmental assessment is actually required (thus providing a screening function for those developments where the requirement for an EA is not clear-cut) and indeed in some countries (for example, The Netherlands) it is mandatory to involve an independent environmental impact assessment (EIA) commission in the process. Early experiences of EIA were that it was applied to a vast array of projects which did not necessarily warrant the level of detailed and comprehensive assessment that was undertaken (Davies 1990). The scoping stage has been highlighted as an area of weakness (Centre for Environmental Management and Planning, 1994; Council for the Protection of Rural England, 1990) while a study of the first 100 environmental assessments undertaken in the UK demonstrated that the poor quality of many could be traced back to an inadequate scoping exercise (Wood and Jones, 1991). A more recent study (Glasson *et al.*, 1996) undertaken for the DoE identified that the standard of scoping exercises has since improved, however, while others concluded that early consultation and scoping of ESs should be mandatory (ECC, 1996)

Many of the environmental statements submitted so far have been subject to criticisms when exposed to critical review and analysis, with numerous and various inadequacies being noted (Wood and Jones, 1991; Coles and Tarling, 1991) which can be blamed largely on poor or inappropriate scoping.

There have been several definitions of scoping as a process and an activity (USAEC, 1973; States *et al.*, 1978; Fritz *et al.*, 1980); however, one of the clearest, most user-friendly (and shortest) for the non-specialist is:

> a very early exercise in an EIA in which an attempt is made to identify the attributes of components of the environment for which there is public (and professional) concern and upon which the EIA should be focused
>
> (Beanlands and Duinker, 1983)

This definition is interestingly similar in its aims and clarity to that used in the United States of America Regulations on environmental assessment (Regulations for implementing the procedured provisions of the National Environmental Policy Act, 1986).

The importance of scoping

Early evaluation of the purpose, need, alternatives and assessment of a proposed project is important to identify the direction of the environmental assessment. This has been summarised (Carpenter and Managers, 1989) as the 'Seven Ws':

- Why is the project proposed?
- What will the final project look like? What is the magnitude of the project? What design options are available?
- When is the proposal to be implemented? When will the construction phase/operation/decommissioning (if applicable) take place?
- Where are the preferred sites for the proposal? What are the possible alternative sites?
- Who is the project proponent? Who will build and operate it?
- Which public bodies or communities will be interested or affected by the project?
- How (*sic*) will the project be implemented?

By addressing these issues a clear indication will be provided of where the project is heading, which can be used as the basis of initial scoping studies/meetings. The World Bank guidelines on environmental assessment (World Bank, 1991) place great emphasis on identifying environmental issues early in the project life, thus enabling environmental improvements to be incorporated into the design of projects, as well as avoiding, mitigating or compensating for adverse impacts, and therefore avoiding costs and delays in implementation.

The importance of scoping in environmental assessment can be gauged, for example, by the publication by the Environment Agency of a specific scoping handbook for water-related developments (Environment Agency, 1996). This document, in addition to over 60 Scoping Guidance Notes for different types of development and activities (from reservoirs to bait digging) was prepared by the National Rivers Authority before the amalgamation of the regulatory authorities in 1996 and is likely to be followed by similar documents from the Environment Agency to reflect those impacts that may occur on the land and to the atmosphere as a result of specific developments. At an Oxford Brookes University forum on the implications of the amended EIA Directive there was unanimous support for the introduction of mandatory scoping, with the onus placed on the developer to prepare a draft scoping document for review by the planning authority and other consultees (Oxford Brookes University, 1997).

In the UK, scoping tends to be an informal process that is often carried out either by the proponent of the scheme, where they have sufficient expertise, or externally by consultants offering such services. Local authorities are increasingly undertaking such studies (partially or in totality) either on behalf of the proponent or on a parallel route to that being conducted by the developers (or their consultant). Comparison of the results of the two studies can often reveal important insights to the items upon which focus should be concentrated in any subsequent environmental impact assessment,

Scoping can be considered as a means of defining a scope of work; for a consultant undertaking the study this can be critical as a means of agreeing their input with the client (the proponent). It will be of prime importance for consultants to identify the boundaries of the study. For the proponent, the scoping study can identify what studies will be required, how long they are likely to take and whether he has the resources to manage the study. It is interesting to note that one of the 1997 amendments to the EIA Directive may force the widening of the scope of work for some types of development. The cumulative impacts of a road scheme may be required to be assessed and the impact from associated developments (such as overhead power lines from a power station) may need to be studied, where currently they are sometimes ignored or are the subject of an 'unrelated' environmental statement!

Approach

Scoping can be undertaken in a variety of ways; scoping and environmental assessment are inextricably linked and it is becoming increasingly rare that the latter is undertaken without the former. Scoping studies, however, are often undertaken without the subsequent environmental assessment being completed. This can occur as a result of concerns over the success of the application where it to be submitted in its current format or without significant changes (such as process modifications or site relocation). Similarly, as the result of the scoping exercise an environmental assessment may not be deemed necessary. This indicates the value of undertaking scoping studies; issues of concern or importance can be identified at an early stage, rather than after submission of the planning application and environmental statement.

Regardless of the scale of the development being considered, environmental statements are invariably lengthy documents (often running to several volumes) that have been compiled over a long time period at considerable expense; rejection of the environmental statement at the first hurdle, because inappropriate methodologies have been used or the type of development is in conflict with (for example), the intended land use of the site, would be an embarrassment at best and a significant and costly error at worst. Scoping can reduce the possibility of this occurring.

Increasingly it is becoming normal practice for potential developers to invite proposals for scoping studies to be undertaken before the commissioning of the environmental assessment. Such studies, when undertaken, give the developer an insight into the issues that will require study and identify any potential problems with the development proposals in terms of the site layout or conflicting land uses, and commit only a small proportion of time and resources to the project (relative to the instruction to undertake a full EIA, which would be significantly larger in terms of both man-hours and fee).

This approach to the scoping enables the correct focus to be placed on the proposals, as well as assembly of the correct project team at the start of the environmental assessment process. Identifying, for example, a specialist in entomology who can provide the necessary input to a programme of work that has been under way for some time can be difficult at best, and costly or impossible within an achievable timescale at worst. For the developer, a project specification can be prepared from a scoping study and this in turn can be used as the basis for competitive tendering among consultants for the undertaking of the environmental assessment. It is important, if this route is selected, to encourage the tenderers to use their professional experience and expertise to refine the scope of work where possible and appropriate. The consultants adopting such an approach should make clear their reasons and justification for so doing.

The main purpose of the scoping study, however, is to identify precisely what issues will require inclusion in the environmental statement, and what methodologies are appropriate for undertaking the assessment of impacts, and to provide guidance on the depth of assessment that will be required. There is no legal requirement under current regulations for the results of the scoping study to be reviewed by the local planning authority.

In the UK the planning authority has the power to require additional information to be submitted or, where relevant, evidence to be supplied to verify the findings of the environmental statement. The planning authorities will, when requested, more often than not provide guidance on the required content of the environmental statement and indicate what information they will be expecting the document to contain. Although it should be noted that it has been estimated that one-fifth of all local planning authorities (LPAs) have no experience of the assessment of environmental statement quality (Petts, 1996), it is reasonably safe to assume that it is unlikely that the same LPAs would not be in a position to recommend strategies or areas for assessment within the environmental statement.

Where a scoping study has been undertaken that has involved consultation with the regulatory authorities and the advice provided has been followed during the assessment, it is unlikely that on submission of the finalised environmental statement further information would be required to be submitted. Similarly, where the scoping study has identified that a particular issue does not require study and exclusions are agreed, the transparency of the final ES should be aided by stating the reasons for this and with whom the agreement was made. Such an approach is significantly preferable to the issue appearing to have been ignored. Such an omission has been identified as a major pitfall in the completion of statements (Petts, 1996).

In the event of submission of additional information to the planning authority being required, this can often cause considerable delays in the determination of the application; delays to the award of planning permission can significantly affect the timing of the development and the delicate financial cash-flow balance that accompanies many smaller developments. The lack of information in a particular area of a environmental statement may often not be noted for several weeks and not communicated to the proponent for several more. In the meantime the project team responsible for the preparation of the document may have been disbanded and become committed to other studies. Such situations are therefore best avoided; this can be achieved through an effective scoping exercise.

Poor scoping can also affect the chances of a successful planning application by failing to identify deficiencies in development design or the inability of the selected site to accommodate the facility with safety (Petts and Eduljee, 1994). Developers constrained by financial costs will almost always aim to squeeze the development onto the smallest possible development site (particularly for infill developments within industrial areas); such an approach may be successfully opposed by either the Health and Safety Executive or the Fire Services on a variety of safety-related grounds.

A similar result may occur through failure to include potential or actual issues of significance from the environmental statement, for example the lack of awareness on the potential for archaeological remains being present on a site could be instrumental in the rejection of planning permission until such information can be supplied to, and reviewed by, the planning authority. The time taken by a return to site to collect such data and by their subsequent interpretation will seriously delay the determination of the application.

There is frequently justification for reviewing the scope of work during the undertaking of the impact assessment process; during the data collection exercises (which are reviewed below) and the impact prediction phases, information may come to light that requires a change in focus or more usually an expansion of the original brief. It is often the case that additional data, that could significantly affect the success of the planning application, are identified that were not identified during the scoping study. Throughout the project, the data and their status should be evaluated and the environmental assessment modified accordingly. Ignoring the issues encountered during the assessment process is an unprofessional approach and one that is likely to create inaccuracies and potentially lengthy delays following submission of the environmental statement.

Methodologies

The methodology adopted for undertaking the scoping study will vary depending upon the probable sensitivity of the proposed development and its scale. The larger and more potentially contentious the development, the greater the need to scope the study accurately and the more complex the methodologies will be to achieve the exercise. If the development being considered is significant (in terms of geographical size and polluting potential), it may be necessary to follow a number of different techniques at the same time and compare the results obtained.

The significance of a potential impact is clearly difficult to quantify or assess accurately before the assessment process itself has been undertaken. As a consequence, estimation of significance should be attempted during the scoping study through comparison with the perceived effects that may occur on a variety of receptors (defined as components of the natural or man-made environment, such as air, water, a building or a plant, that are affected by an impact) (DoE Planning Research Programme, 1996) and their relative importance in the area being considered. This is where the experience and knowledge of a suitable specialist will be extremely beneficial. The impact of the erection of a radio communication mast, for example, on the ecology of an area will be greater in an area designated as a site of special scientific interest (SSSI) than on an existing industrial site (although it should be noted that in different locations other impacts may replace ecology as the important aspect for consideration).

Techniques for undertaking scoping studies have been reviewed (Petts and Eduljee, 1994) and include the preparation of checklists, matrices, networks and cause–effect diagrams. All have advantages and disadvantages and some are more suitable than others, depending upon the type of development and who is undertaking the exercise (Leopold *et al.*, 1971; Sorenson, 1971; Dee *et al.*, 1973; Clark *et al.* 1981; Andreottola *et al.*, 1989; Kletz, 1992). Each technique requires a varying level of experience and knowledge of the development itself and the type of impacts that are likely to occur. An example impact matrix prepared as part of a scoping study for a landfill proposal is presented in Table 9. An example of a checklist is included in published Government guidance (DoE, 1994b).

Timing

The timing of the preparation of the scoping study will vary but almost always it should be undertaken at the beginning of the project, or ideally some time before the environmental assessment has commenced in earnest.

Table 9. Simplified impact matrix[a] for a proposed china-clay disposal site (MRM Partnership, 1992)

Issue	Construction phase	Operation phase	Post restoration
Air	×	×	—
Noise	×××	××	—
Water	××	×	√
Ecology	×	××	√
Archaeology	—	—	√
Transportation	×	××	—
Socio-economic	√	√	√
Landscape/visual	×	××	×

[a] Key:

—	No significant positive or negative impacts
√	Minor positive impacts
√√√	Major positive impacts
×	Minor negative impacts
×××	Major negative impacts

This may be weeks, months or even years before the assessment process begins; however, if the gap between the two is greater than a few months it may be necessary to update the scoping study to take account of variations in environmental conditions at the site being considered.

The role of consultation in scoping

The general role of consultation in environmental impact assessment is discussed in a later section, but it can provide a very valuable tool during the scoping exercise and some discussion on this issue is appropriate here. There are several forms of consultation requiring differing levels of input and effort (by the proponent, consultant and LPA). The preparation and issue of a letter that outlines the intended or proposed development (including details of the proponent, the site location and activities that will take place), and requests initial comment from the statutory and non-statutory consultees can often result in valuable and constructive initial comments being received.

The Government recommends that the planning authority and statutory consultees are invited to participate in the drafting of the terms of reference for an environmental assessment (DoE, 1995a). This is also an excellent means by which to gauge the reaction to the development, and how the planning application may be received on its submission.

109

The local authorities often appreciate and encourage informal involvement from the beginning of an impact assessment and can be supportive through the provision of relevant data, the identification of specific issues of local concern and the supply of contact names of individuals and organisations who may have either specific concerns or access to useful data.

Other methodologies that can be adopted as part of the scoping study include questionnaires and surveys (either on an informal basis where the specific development being considered is not identified, or on a more formal basis that includes identification of the development and its proposed location). This method of scoping is considered less desirable, as its effectiveness requires a significant amount of planning while the interpretation of results is often open to considerable professional disagreement.

Scoping studies are almost always subject to time constraints; this invariably precludes the use of widespread consultation with all the interested parties and members of the public. Consultation with the public can be a useful exercise but it can be very time-consuming and fraught with difficulties unless considerable preparation is made beforehand. An approach sometimes taken is that small community meetings are held to discuss the proposal before formal EIA hearings; residents and interested parties attend the meeting and are given the opportunity to discuss their concerns in the presence of the assessment team and also representatives of the regulatory authorities. The interchange that is reported is non-confrontational and the participants are not required to defend their positions. While such meetings could improve the understanding of the impacts that may occur as well as providing the assessment team with valuable information on the real and perceived issues of concern associated with the proposals, they are not always successful as the proceedings may be instrumental in mobilising opposition, who may dominate them, before the assessment procedure has even begun.

Types of impact

The types of impact that will occur as a result of a particular development depend upon:

- the type of development;
- the location of the proposed development;
- the characteristics of the site and the surrounding environment;

- methodologies to be utilised during the construction phase; and
- the nature of activities taking place during operation of the facility.

As identified in Chapter 4, environmental assessments that are undertaken require impacts to be identified in terms of their effects on

- human beings,
- flora,
- fauna,
- soil,
- water,
- air,
- climate,
- the landscape,
- the interaction of any of the above,
- material assets, and
- the cultural heritage.

During the preparation of the environmental statement it will be necessary to describe the impacts in terms of whether they are direct or indirect, whether the identified impacts will be of short, medium or long term, whether they are of a beneficial or adverse nature, whether they can be reversed and whether they are cumulative. It will also be necessary to describe the type and extent of impact after mitigation measures have been taken into account to reduce the scale or nature of the impact, i.e. the residual impact.

It is also necessary to describe the impacts in terms of their significance, both in the geographical sense (i.e. are impacts likely to be felt locally, regionally, nationally or globally?) and also in terms of the time before the impacts are realised (i.e. are the impacts likely to be acute or chronic?). Acute impacts most often occur with respect to potential release of chemicals such as the discharge of emissions from a stack or the accidental release of chemicals from a tank into a river.

When will impacts occur?

The periods when environmental impacts will occur also need to be clearly identified and discussed. This will include impacts during the following three stages:

- construction;
- operation;
- closure/Post-Operation.

111

The construction phase

The commencement of construction activities at a development site may be the most noticeable of all the phases as the site may change from a green-field site to one with a radically disturbed and different appearance. Some types of development, however, will be barely noticeable during the construction phase; the development of an existing quarry site as a landfill, for example, will require some construction-related traffic (as the leachate and gas control systems are installed), but more probably it will not be possible to differentiate from the activities taking place during the removal of mineral or aggregate that may still be occurring.

Construction work associated with many types of development has the potential to cause great disruption to the local environment due to the large number of items of plant and equipment that may be required. Activities during this phase can vary in intensity but the potential is high for environmental impacts to be greater during this phase than following completion of the development and during its operation.

The type of construction activity will vary according to the scale and type of development; Table 10 provides some general guidance on the activities that can be expected in most construction programmes and the impacts that may occur. Considerable guidance is also available for specific construction activities (CIRIA, 1993a).

The potential for reducing or mitigating the impact from construction activities will depend upon the proximity of sensitive receivers and the magnitude of the impact; this is discussed in greater depth in subsequent sections but it is relevant to indicate at this point that reducing the intensity of activities will invariably increase the time during which construction related activities will occur.

The operation phase

All types of development will have an impact of some form or another during their operational phase. The scale, significance and type of impact will of course vary considerably between development types and the locations where they occur and are significant. Some developments will be relatively benign in nature and will not produce large amounts of liquid, effluent or noise emissions for example, but they may have less specific and more subjective impacts in the form of visual and landscape concerns.

For industrial developments most concern is often expressed over the impacts that the facility will have during the operational phase. These invariably occur in the form of various discharges to air and water, occur either intentional as a result of the need to dispose of an industrial by-product or accidental through the failure of an in-place system that

Table 10. Examples of construction activities and consequent environmental impacts

Activity	Special operations	Main potential environmental impacts
Demolition and site clearance	Removal of on-site buildings and structures, demolition and site preparation of temporary access roads, removal of topsoil, clearance of vegetation, removal of existing services. Erection of site offices	Noise, dust, ecology, vibration, water quality, waste disposal, visual, traffic
Tunnelling	Boring, blasting and spoil disposal	Noise, vibration, waste disposal
Earthworks	Dredging, excavation for foundations and services, re-profiling site contours	Noise, dust, landscape, water quality, visual
Piling		Noise, vibration
Superstructure	Erection of scaffolding, erection of build shell, casting of structures on site, wall construction	Visual, noise
Roads and paving	Access road, junctions, public highways, car parks	Traffic, noise, landscape, vibration
Fitting out	Installation of plant and internal services	Traffic, noise, visual
Landscaping	Replacement of topsoil, planting, seeding	Noise, air, traffic, visual, landscape

has been incorporated as part of the design of the plant to guard against such releases.

It is often easier to envisage and predict the impacts from an industrial facility than from other types of development. The type and volume of emission, effluent or solid waste produced by the factory can be predicted and will determine the potential scale and significance of the impact; for example, the release of untreated effluent from a metal plating works will be of significantly more importance than the venting of uncontaminated steam to the atmosphere. Each will require assessment (and examination of the risk that will occur), as the scale of impact will need to be assessed. The methodologies for predicting impacts are discussed later in this section.

The closure/post-operation phase

Many environmental statements do not address the impact of the development when it ceases operation, becomes closed and is demolished. This is simply because, at the time of preparation, the end-life of the development cannot be envisaged. For example, the preparation of an environmental assessment for a road or marina development may include reference to maintenance programmes that will maintain the property or development in the 'as-built' condition. There are circumstances, however, that will eventually result in this not being the case and an environmental impact will occur as a result of the demolition or closure programme (in 5, 50 or 500 years). Redevelopment of the site for the same or another purpose, however, may be required to address the impacts that will occur during this phase. Thus, the loop will eventually return to the construction phase (see above); this is increasingly likely as more and more developments become subject to either formal environmental assessment studies or partial studies (such as specific noise or traffic studies).

One notable exception to this is the development of landfill sites where consideration of the restoration methodology, final land use, risk control measures, landfill gas and leachate management and landscape management proposals will represent a significant part of the environmental statement. The reason for this apparently different approach lies in the very readily identifiable impacts that will occur if landfill management is not undertaken correctly; these impacts will become manifested in the form of off-site landfill gas and leachate migration.

The introduction of requirements for a landfill site operator to be responsible for the site for a period of time after receipt of the last waste delivery at the site (under the Waste Management Licensing Regulations 1994) ensures an involvement by the operator and identifies firmly placed responsibility for the site. A landfill development could therefore be considered as having two operational states: firstly for the receipt of waste, and secondly for the organic degradation of that waste following closure of the site. Both will require assessment in the environmental statement.

What impacts will occur?

The issues that will require inclusion within the environmental assessment and examples of how and where impacts may manifest themselves as a result of a development taking place are reviewed below. This is intended only as a guide to the type of impacts that may occur and should not be taken as comprehensive in its extent or coverage.

Water resources

There are two main spheres of impact by any major development on water resources; effects on surface waters and impacts on groundwater. Impacts can occur not only in the form of pollution of both types of water body but also as a result of abstraction proposals which may affect the availability of supply or take the form of flooding.

The discharge (accidental or intentional) of pollutants through the generation of effluents or leaks/spillages to water bodies can affect the quality of the receiving water body. During the preparation of an environmental assessment, it will be necessary to quantify, in terms of effluent strength and volume, the nature of the intended discharge, its location and whether the discharge is of a constant or batch release. Similarly, the potential for accidental releases on the site should be assessed. It will then be necessary to review the potential impact that the release may have and the impacts that may occur on water quality, downstream extraction points and aquatic life. Discharges to surface waters are permitted under the Water Resources Act 1991 but they are subject to limits on effluent strength and volume and an appropriate licence must be obtained from the Environment Agency (or SEPA in Scotland) before any discharge takes place. In the event of a discharge not being permitted, disposal of the effluent to the sewerage system or to a specialist treatment plant can be considered, also subject to relevant legislation such as the Water Industry Act 1991 and Trade Effluents (Prescribed Processes) Act 1989/90/92, although the capacity of any existing receiving system must also be addressed.

The impact of developments on groundwater resources can alter the balance between stream flow and groundwater. Groundwater supply can be diminished where rainfall is intercepted and discharged directly to stormwater drains, although interception over very large areas would be required for there to be a noticeable effect on the groundwater resource (and for most developments isolated from others, this will not occur). Sensitive habitats (such as upland mires) can be dependent upon high water tables and even individual trees can be affected by falling water tables. Protection of aquifers and their recharge points from contamination attracts a high degree of attention by the regulatory authorities, who have prepared aquifer protection policies that must be consulted before any discharge to surface waters is considered.

115

The abstraction of water can have significant impacts on both surface water and groundwater resources, with indirect and consequent impacts on ecology. The hydrological characteristics of a site may also place constraints on a development, particularly in terms of the risk of flooding or erosion.

The key water-related issues for consideration in any impact assessment may be summarised as:

- water quality;
- water body levels;
- flooding;
- existing uses;
- supply and availability;
- effluents — type, strength, volume etc.;
- accidental releases;
- aquifer protection/recharge;
- bank erosion;
- ecology;
- biological characteristics; and
- chemical characteristics.

Air pollution

The types of air pollutant that may be produced as a result of a development are listed in Table 10.

Although a combination of these pollutants may be produced by any one facility, there are now stringent controls on the concentrations and amount (volume) that are permitted to be discharged from a point source such as a factory (under the Environmental Protection Act, Part 1, 1990) and as a consequence most significant emissions of air pollutants are first directed through gas-cleaning systems that remove large amounts of the pollutants present in the discharge.

Non-point-source (or linear) emissions include vehicles (which in turn represent single points of discharge that are mobile) so they represent one type of impact that will result from the construction and use of a road, are potentially harder to control and to mitigate; if air quality impacts are predicted, efforts must be concentrated on selecting the most appropriate route.

Some of the key issues to be taken into account are:

- the type of pollutants (including combinations);
- the concentration of pollutants;
- the volume and rate of discharge;

Table 11. Examples of air pollutants and their sources

Pollutant category	Description/examples	Sources
Particulates	Emissions of particulate matter in the submicron and micron size ranges which remain suspended in the atmosphere	Construction activities, coal combustion, mineral workings, cement plant, industrial processes
Acid gases	Oxides of sulphur and nitrogen	Products of combustion e.g. power generation, transport Chemical processes, waste incineration
Hydrocarbons	Solvents and fuels	Fugitive emissions from storage areas
	Unburned fuel due to incomplete combustion	Motor vehicle emissions
Ozone/oxidants	Secondary pollutants	Action of sunlight on atmospheric aerosols polluted with hydrocarbons and nitrogen oxides
Trace metals	Particulate matter containing toxic trace metals	Elevated levels may occur in process streams, e.g. from incinerators or lead works
Soots, dust, grit	Particulate matter above the respirable size range and typically 70–80μm in diameter	A wide variety of industrial processes, building works and transportation activities
Carbon monoxide	Toxic gas	Incomplete combustion of fuel
Carbon dioxide	Combustion of carbon compounds (oils, coal, etc.)	

- whether there are point or area sources;
- any regulation of emission;
- the variation of emission;
- the speed and height of discharge point;
- proximity to residential areas;
- the topography;
- whether there are other unrelated emissions nearby.

Ecology

The study of ecological impacts may require an examination of many groups of species including vascular and non-vascular plants, mammals, birds, invertebrates, fungi, amphibians and reptiles. The interrelationship between these species in their environment, i.e. the ecosystem, can be very complex and all sites will be different. Each species will be dependent upon its habitat and therefore habitat conservation is necessary to ensure species conservation and diversity

The demand for protection of ecological resources is gaining in significance and represents many individual's concern over new developments where encroachment onto green field sites is scheduled to occur; emotive reaction (justified or not) to habitat destruction can create significant delays to the development process in the event of the environmental assessment process not being completed in sufficient depth. One recent study of 37 environmental statements relating to road developments observed that they were significantly lacking in detail (Treweek *et al.*, 1993); indeed, five statements made no mention whatsoever of any potential ecological impact.

A wide range of ecological issues may require study in an assessment but they will be dependent upon the location of the site and the type of habitats that can be found in that area; ecology is a complex and very broad discipline that covers plants and animals in a variety of different ecosystems (e.g. terrestrial, freshwater, estuarine and marine). Impacts that may occur are even more difficult to define because of the often very complex interrelationships between them, i.e. the effect of communities on other communities.

The aim, function and role of the ecological component within the environmental assessment has been defined (Morris, 1996) and should be to assess the conservation value of a species within the impact area of a development (which may include habitats and species outside the immediate site boundary of the development), to identify the likely impacts of the development, and to consider and propose mitigation measures that can be taken to avoid, reduce or minimise the impacts. The broad aim should therefore be to maintain biodiversity. While it may appear an irrelevance for many proposed industrial sites to consider this as a constraint (because of the apparent lack of any ecology on the site or due to the relatively small land take at a green-field site) it is an important concept that should be adopted in all environmental assessments; 50% of native broad-leaved woodlands and lowland heaths have disappeared in the 40 years leading up to 1990 (Harding, 1992). Increasingly, more local planning authorities are looking to ensure that the issue of habitat replacement has been addressed in the environmental assessment and that there is a commitment from the developers to contribute to, for example, wildlife corridors.

Protection of habitats and flora/fauna in the UK is now achieved through the EC Directive on environmental impacts (EU Directive 85/337/EEC Assessment of the effects of certain public and private projects on the environment) and the implementation of this by the environmental assessment regulations and other more specific legislation (such as the Wildlife and Countryside Act 1981 (amended 1985 and 1991) and the Protection of Badgers Act 1992). Prevention of habitat and animal destruction is afforded through protection of specific areas (such as National Parks and sites of special scientific interest) and through protection of specific species such as badgers (Protection of Badgers Act 1992). However, designation of sites in this manner does not guarantee their protection. Many developments are permitted within National Parks and there are numerous cases of developments taking place within SSSIs (over one-sixth of top nature reserves were at risk according to the Friends of the Earth in 1995). The requirement for development for economic and socio-economic reasons often therefore results in a clash with the desired ecological protection of an area. The environmental assessment process should allow the impacts on ecology to be studied and rational decisions to be made on the relative merits of each and the mitigation measures that can be adopted to reduce the scale of impact.

In summary, the ecological issues to be considered are:

- habitat destruction;
- biodiversity reduction;
- habitat disturbance;
- community stability;
- species destruction;
- severance of migration routes;
- reproduction viability; and
- maintenance of the food chain.

Agriculture

Land quality and the value of land for a variety of agricultural purposes will be required to be assessed where agricultural land take is a potential issue as a result of a proposed development. All agricultural land has been classified by the Ministry of Agriculture, Fisheries and Food into one of five grades (where Grade 1 is the best and Grade 5 is the poorest) according to the degree to which permanent physical characteristics impose limits on it for food production. The Government has assigned differing levels of protection to these different land types (DoE 16/87 Development involving Agricultural Land) and a land classification survey would be required as part of an environmental assessment to identify the type and value of land that may be affected. This is necessary in order to control the rate at which land is taken for development purposes.

The commercial viability of a piece of land may also be required to be assessed particularly if the development results in farm severance and fragmentation, or results in difficulty in accessing land plots. As demand for some agricultural products fluctuates over time and it is unlikely that the land will be returned to agricultural purposes following the development (except in the case of mineral and landfill type developments), it is a general and unstated rule that the best and most versatile land should only be built upon in the event of no other site being suitable or available for use.

Geology
Assessment of geological impacts within an environmental assessment mostly tends to centre on the potential presence of contaminated land at the site and the possible impacts this may have on the health and safety of construction workers or the final end-users (e.g. in a residential development with gardens/allotments). Consideration of the geology of a site will be critical for addressing the impacts on the civil engineering methodologies that may be required at a site and the potential, for example, for the strata to satisfy the load-bearing requirements for the development being considered.

The development of sites in coal-mining areas will also require careful assessment because of the subsidence problems that may be associated with them and also (as a related, but separate issue), the potential for mine-gas migration and accumulation.

The presence of specific and unusual geological formations at the surface will also be required to be identified and the impact upon them assessed where relevant. Some features are protected through SSSI designations, which are often associated with specific ecological habitats.

Issues of concern relating to geological resources that may be impacted are:

- mineral resources;
- geotechnical stability;
- geomorphology;
- topography; and
- soil

o   profile,
o   physical characteristics,
o   erosion,
o   chemical characteristics and
o   pollution.

Noise
Noise can have a significant effect on the environment and on the quality of life of individuals and communities. Noise is of growing importance within all communities and is of concern to all those affected by it; this is demonstrated by the Government's increasing willingness to legislate against noise pollution (for example, Environment Protection (Statutory Nuisances) Act 1990 and Noise and Statutory Nuisance Act 1993). Daytime noise levels in the UK are higher than the levels desirable to prevent significant annoyance at about half of the dwellings in England and Wales (DoE, 1992).

Almost all types of development will produce noise impacts, either during the construction phase or as a result of the operational activities (or during both); the significance of the impact will be determined by the extent of nuisance caused. At the acute stage, noise can cause permanent damage through deafness and hearing loss, while at the chronic level it can case nuisance, stress and loss of privacy. The impacts may be indirect in that a development itself will be relatively noise-free (either because they are fairly benign or because attenuation measures have reduced the level of noise), but the vehicles attracted to the development may create a significant rise in traffic noise.

Noise is measured in decibels (dB), which operate on a logarithmic scale: therefore a 10 dB increase in noise levels subjectively represents a doubling of loudness and the addition of two sounds at the same level leads to an increase of 3 dB rather than a doubling of the decibel rating. Noise consists of changes in pressure variations which are measured in terms of frequency and amplitude. The frequency of a noise or sound is measured in cycles per second (hertz; Hz), and depends on the length and velocity of the sound waves, while the amplitude represents the height of those waves and is measured in units of force per unit area (N/m^2).

There are a number of factors that influence the significance of the impact of a noise source. These include distance attenuation (sound is reduced with distance from source and by scattering and absorption) and the presence of barriers (such as a wall surrounding the source or double glazing to protect the receptor) which will reduce received levels

121

noise by reflection and absorption. Other aspects that can influence noise levels include absorption by soft ground (grass, trees etc.), topography and atmospheric conditions (e.g. wind will carry sound in the direction in which it is blowing).

The type of noise and its frequency are also important aspects to address. Engine and exhaust silencers operating on a landfill, for example, reduce noise levels from all frequencies except the low ones, which can cause considerable nuisance. Similarly, the duration of the sound activity (in terms of minutes or hours per day) and whether it is intermittent or continuous (for example, reversing bleepers can be particularly intrusive) greatly influence the degree of annoyance that can be caused.

A checklist of issues to be taken into account when examining potential noise impacts is as follows:

- Level?
- Frequency?
- Timing?
- Character?
- Periodicity?
- Intensity?
- Distance?
- Prevailing wind direction?
- Regularity?

Landscape and visual impacts
Concerns from the public over a specific development proposal are often manifested most directly as a result of the impacts (real or perceived) on the landscape. Such impacts are therefore often the focus of most popular concern.

The impacts that can occur are site-specific in nature, broad, wide-ranging and often subjective; this can lead to difficulties in the prediction of impact scale and significance.

The Institute of Environmental Assessment (IEA) provides a useful distinction between landscape and visual impacts; landscape impacts result in changes in the fabric, character and quality of the landscape, whereas visual impacts relate to the changes in the appearance of the landscape and the effects of those changes on people (IEA and Landscape Institute, 1995).

The relationship between landscape and visual impacts is often difficult to separate clearly because of the overlapping nature of the two terms. Figure 3 identifies the relationship.

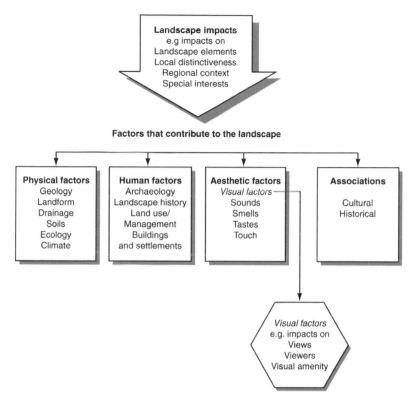

Fig. 3. Relationship between landscape and visual impacts (after IEA and Landscape Institute, 1995)

Landscape and visual impacts are likely to be the most subjective element of the environmental assessment but, as will be illustrated in subsequent sections, techniques are now evolving that reduce the subjectivity of the assessment; indeed, standard approaches to this element of an assessment are now encouraged and utilised (IEA and Landscape Institute, 1995).

For some schemes, however, the importance of this element within environmental assessment cannot be understated. Changes to the landscape character can often occur, and concern over what a development will look like will often promote an emotive response, especially for developments at green-field sites. In such cases the development may not be seen in isolation from others developments but should be considered as part of wider plans for development of adjacent sites. Conversely, in derelict areas, the redevelopment of a site will result in a very positive impact on the landscape and visual environment because of the poor relative quality of the existing landscape and the improvements that will occur if the development were to proceed (Rust Environmental, 1995).

The assessment of a particular site will require the identification of the existing landscape character and condition (including consideration of the presence of designated landscape areas or areas of outstanding natural beauty (ANOBs), an evaluation of the views into the site from a variety of locations and distances (static and mobile — for example glimpsed views of the site while travelling along a road) and an examination of how the development may impact upon the site and the surrounding area. This latter aspect may require study of the site from a considerable distance; for example, the construction of a road through a valley may be visible in some places over a considerable distance.

There is frequently a clear overlap of landscape and visual impacts with other elements being considered in the environmental assessment, and each of the elements must be addressed in association with each of the others. Examples are:

- planning — impacts on designated areas of landscape interest;
- ecology — impacts resulting from the removal of habitats;
- air quality — impacts from dust can be mitigated through barrier planting (although the trees/bushes may be impacted upon);
- noise — the construction of earth bunds can reduce noise impacts but cause visual intrusion;
- hydrology — the formation of a water body will significantly affect the visual character of a site.

The importance of each of the issues, and indeed of the whole landscape and visual assessment, varies enormously from area to area depending on the type of landscape and the type and scale of development being considered. Impact assessment is further complicated by a wide range of influences differing from community to community and from individual to individual in terms of which elements are important and which are of secondary importance. Elements that influence an individual's response to a development have been assessed (Goodey, 1996) and include age, gender, activity patterns, family grouping, education, holiday preferences and inherited traditions.

While there is often a clear overlap between landscape and visual impacts, one can occur without the other. For example, the siting of a clinical waste incinerator in Avonmouth had a minimal landscape impact but elements of the development were predicted in the environmental statement to impact on the visual environment (Rust Environmental, 1996a). For all developments, however, it will be necessary to assess the landscape and visual impact that will occur, and to arrive at conclusions as to what the level and scale of impact will be.

Aspects to consider for assessment of landscape/visual impacts is given here as a checklist:

- type of landscape;
- visual obstruction;
- visual intrusion;
- topography;
- landscape use intensity;
- planning controls;
- community influence; and
- landscape value.

Archaeology

Archaeology plays an important role in both leisure-time activities and education as well as being of great importance as part of the UK's tourism industry. Important sites can be very valuable not only in terms of the educational and research value they may represent, but also as a financial asset.

Because of the long history of human settlement in the United Kingdom and the ability of some environments to protect and store remains, there are many areas of the UK that may represent an archaeological resource of some value and importance. The age of those remains and their importance may date back to several hundred thousands of years ago or to almost the present day (for example, in the case of a crashed Second World War fighter aircraft).

The presence of remains may be indistinct to the layman — for example, field boundaries (often covered by thousands of years of other human activity), fragments of pottery or Second World War pillboxes. It has been estimated that there are over 600 000 known archaeological sites in the UK (Bourdillon *et al.*, 1996) and the main focus for concern among County Archaeologists is the rate of development that is taking place (in comparison with the last 500 000 years), and that has been resulting in the destruction of such remains, known and unknown.

The importance of the remains will depend upon a number of factors, summarised as:

- type of remains;
- location;
- rarity;
- condition;

- age;
- potential for excavation; and
- potential for destruction by proposed development.

The presence of remains or an archaeological resource will not necessarily preclude the development from proceeding. Identification of the find, its careful study and documentation to identify its value and significance may be the appropriate way forward initially, rather that its immediate excavation. The latter cannot be ruled out however, until the significance, value and worth of the remains has been assessed correctly. Protection of the resource in its existing position may also be appropriate and often represents a satisfactory method of proceeding.

Traffic

Impacts from a development may occur on the road networks in the neighbourhood and users of those roads; it is true to say that almost all major industrial and commercial developments will require some alteration to the road network in order to provide access for people and for goods. Impacts can occur in a variety of forms, such as the attraction of large numbers of vehicles to a road system that has insufficient capacity to deal with them, or the unintentional routing of HGVs through residential areas. Traffic impact assessment is of great importance and is now often undertaken as a separate and identifiable component of site redevelopment; the results of such studies will normally also be incorporated within the environmental assessment.

Well-designed new road layouts can improve conditions for users, including pedestrians, cyclists and users of public transport, and can relieve congestion, noise and air pollution.

Some of the key aspects likely to be considered in a full traffic assessment for both existing and planned routes as a result of a proposed development are:

- existing traffic numbers (and any likely increase thereof);
- highway link and junction capacity;
- driver delay;
- accident rate;
- traffic speeds;
- road condition;
- site lines;
- pedestrian/cycle flows and access;
- proportion of HGVs;
- sineage.

The study area that will be required to be assessed as part of a traffic study will vary depending upon the size of the development under consideration and the number and character of the traffic that will be generated or attracted as a result. For small developments, the additional traffic flow may represent an insignificant burden to the existing network; then the study can be localised and focused on, for example, site access and turning circles on the site itself. Larger developments, however, may attract such large numbers of vehicles that attention would be required to be focused on routes, capacities and sineage some many miles away from the development itself.

Waste disposal

Proper disposal of waste must take place at all times. The Government has adopted the principle of sustainable waste management; the hierarchy of actions to reduce the impact of waste management proceeds in the following order — wastes should be avoided, reduced, re-used, recovered (including recycling) or as a last option rendered safe prior to ultimate disposal. The Government also subscribes to the proximity principle, under which waste should be disposed of close to the point at which it is generated. Disposal of waste to water and air will have been described in other sections of the environmental statement, and full consideration should also be given to the disposal of solid wastes.

The generation and subsequent disposal of waste is an important issue during both the construction and operational phases. The type and volume of waste produced and the timing of its production (i.e. during which phase it will be produced) will depend upon the nature of the development; for example the construction phase of a road scheme generates more wastes than are produced during its operation, while the construction and operation of a commercial and recreational development produces a variety different types and volumes of waste during the construction and operational phases.

The assessment of waste disposal issues will have close links with other elements of the assessment that require study. Assessment of waste generation rates will be required to assess the volume of traffic that leaves a site, for example. During demolition work or large-scale earth moving this can create a significant impact on the transportation network surrounding the site and will create noise and dust impacts on nearby residential areas or recreational resources.

The impact on the intended waste disposal facility must also be considered; the generation of a large amount of waste may impose a significant burden on a landfill's abilities to receive all the waste within the programme planned. This can be significantly more important in the event of the required disposal of contaminated soil from a redevelopment site; most landfills that are licensed to receive such wastes will have an upper limit imposed per day on how much of a particular waste they can receive. Failure to take this into account can result in delays in the programme, or the requirement to use an alternative site some distance away.

The waste management issues that should be considered may be summarised as:

- type of waste;
- volume;
- variability;
- programme of production;
- planning controls;
- ease of disposal; and
- suitability for re-use, recyclng etc.

Cultural, social and economic impacts
The environmental assessment will be required to assess how the development would effect a locality in terms of economic prosperity, life styles and community impacts (or socio-economic impacts). Again the impacts of a development on the socio-economic factors will depend upon the type and nature of the development proposed. There has been criticism of a 'green bias' in the environmental assessment process; invariably social impacts are under-represented, and any assessment undertaken is lacking in methodology and practice. Tsoskounglou (1997) recently concluded that social impact assessment was a poor relation of the EIA process and needed to be further developed.

The proposed siting of an industrial manufacturing facility in an area of high unemployment will bring not only direct economic benefits to the workforce employed at the site but indirect economic benefits to support industries and expenditure at local shops will rise. Similarly, the construction of a new town close to a small village will change the character of that village for ever; impacts that may be expected include increases in the crime rate, increased community stress and reductions in house prices for properties already existing.

Beyond the most obvious ones, socio-economic impacts can be difficult to predict and measure and may not surface for many years. The period of time the assessment should consider therefore requires careful consideration. Similarly, drawing the boundaries around the physical area that requires assessment is difficult, as well as identifying who will be impacted upon. Nevertheless one may illustrate the socio-economic factors to be taken into account by the following list of examples:

- population density;
- social disturbance;
- community isolation;
- communication routes;
- relocation and housing;
- land loss;
- employment loss;
- employment effects;
- community services/structure;
- cultural disturbance; and
- financial costs.

Establishment of site characteristics

The collection of data reflecting the existing environment or ambient conditions at a site is an essential component of environmental assessment as it represents the baseline conditions from which the subsequent impact assessment is undertaken. It may also present the proponent with an accurate description of the site before his involvement there in the event of unjustified claims of environmental damage during the operational life of the development. The status of ambient data in respect of the intended development and the impacts that may occur is illustrated in Figure 4.

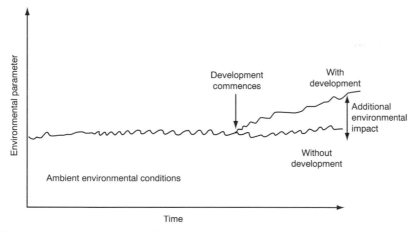

Fig. 4. Ambient data and development impacts

The collection of baseline data for a site is normally one of the first activities undertaken following the completion of the scoping study; indeed it is likely that the scoping study will have identified the data sets that are available, those that are important and critical to the success of the assessment and those that will be required to be collected.

The collection of ambient or background data in a standard and accepted manner will considerably ease the subsequent review of environmental statements; where new or non-traditional methods are used, additional efforts will be required to justify the approach taken and to verify the veracity of the data collected.

The amount of data that will be required to be collected, in terms of the period during which they are collected and the physical number of measurements made, will again be a function of the development type and scale. It is likely that the intended construction and operation of a major power station, for example, will require data collection to be undertaken over a period of years before submission of the planning application and environmental statement, whereas the data collection exercises for a proposed small local waste-transfer station may be complete in a matter of days or weeks.

While data collection is invariably a critical element of the environmental assessment process, it can potentially be extremely costly; to avoid needless expense, the collection of quantitative data therefore requires careful planning and focused objectives. The cost of a recent audit monitoring exercise was reported to outweigh the capital cost of stream-bed and profile improvement works (Brookes, 1997).

Given the importance of the data collection exercise as part of the preparation of the description of the existing environment, it is necessary that the data collected and used are relevant, accurate, reproducible and representative of the ambient conditions. For example, at residential properties facing a road, noise levels during the rush hour are not representative of night-time conditions when noise disturbance may occur; it would be necessary in this instance to undertake monitoring at night in order to obtain relevant data for this period. Similarly, the undertaking of an ecological survey in mid-winter is likely to result in considerable inaccuracies in the data obtained.

The data to be used as part of the assessment should also have been recently obtained. Although data sets collected some years beforehand may be representative of the current on-site conditions, suspicion always surrounds such data and unless their current validity can be verified then problems of confidence will occur (it is often only possible to confirm that the 'old' data are still valid by undertaking additional data-collection exercises).

Data for the environmental assessment are normally obtained either by using those already existing and available, or by collecting new data. A variety of sources can supply data on a particular area. Local authorities, government departments, regulatory authorities (such as the Environment Agency), local and district plans, local interest groups and maps for example may represent valuable sources for the data collection exercise. The type, nature and extent of available data will vary between areas, however, and their validity will need to be assessed before reliance can be placed upon them. Other environmental statements already completed and available for public scrutiny can also provide a very valuable source of data if the document addresses an area that covers the site being examined.

Under the Environmental Information Regulations 1992 (SI 1992 No. 3240) enquiries can be made of the local authorities on whether they hold any information relating to specific sites. The Regulations require that relevant persons holding past or present information relating to the environment should make that information available to anyone who requests it, within two months, at a charge of no more than is reasonably attributable to the supply of the information. Any grounds for refusal to provide the information must be given.

Thus the Regulations may allow access to information held regarding:

- the physical/biological state of any environmental media at any time;
- conditions in and around man-made structures;
- living and dead organisms;
- UK, EC or world-wide matters;
- human health issues;
- activities having a potentially damaging impact; and
- steps/activities taken with the aim of improving the environment.

Primarily this information will be in the form of:

- IPC/LAAPC (Integrated Pollution Control/Local Authority Air Pollution Control) registers of applications, authorisations, conditions, requests, variations and certain monitoring data;
- local authority records and registers kept as part of the contaminated land regime;
- details of emissions, incidents, etc., *required* to be furnished to the HMIP/EA/local authority;
- information compiled by the EA and local authority inspectors;

131

- EA (environmental assessment) registers of applications, licences, authorities, consents, conditions, variations and monitoring and sampling data;
- sewerage undertakers' registers of discharge consents;
- HSE (Health and Safety Executive) reports, investigations, etc.;
- environmental statements;
- waste regulators' records relating to waste management licences and waste carriers;
- local authority information and registers relating to statutory nuisance, especially noise, litter, sewer and drainage maps, etc.

In practice it can be difficult to enforce the Regulations. The exemptions are wide and easily arguable; debates range about whether certain types of organisation are caught by the Regulations at all, information is occasionally presented in a form which is meaningless to the lay person or non-specialist and the cost may be prohibitive.

Examples of other data that may exist and may be utilised (and which may have been identified through the scoping study) include (but are not be restricted to):

- water quality data (held by the Environment Agency),
- information concerning archaeological sites held on the Sites and Monuments Record (held by the County or District Archaeological Officer),
- air quality monitoring data (held either by the District or County environmental health departments or as part of the National Survey),
- structure and local plans illustrating landscape designations,
- information on the geology of an area including information on both surface deposits and underlying geology (available at most libraries), either in map form or as a result of site investigations undertaken on nearby sites,
- traffic statistics, including design capacity, peak traffic flows and accident black spots,
- noise monitoring data held by the environmental health departments as a result of concern made over elevated noise levels,
- ecological data held by English Nature, local libraries and ecological trusts as a result of research undertaken or in the form of information attached to the designation of the site (such as the information held on SSSI designations).

In the event of further data collection being identified as necessary, the scoping study (reviewed above), will play an important role in identifying what data are to be collected (type, number, locations etc.).

For example, noise may be identified as an issue of concern in relation to a particular proposal and then monitoring at a prescribed number of residential properties may be required. Similarly, the scoping study may identify a site as of little or no ecological value (as advised by English Nature) and then a series of detailed ecological studies would not be necessary.

The objectives of each aspect of the monitoring and data collection exercise should be fully understood and the programme designed accordingly to meet those objectives. A common complaint of submitted environmental statements is the seeming lack of knowledge as to why the data have been collected or what to do with them once collated. In some studies the aim has been to collect a small amount of data on all aspects; this results in superficial surveys and there is frequently the requirement to undertake the survey more accurately at a later date. Individuals or companies should undertake data collection only for the areas for which they are qualified. This can be particularly important in ecological surveys and the competence of those involved must be assessed before the studies are conducted.

It is necessary in the environmental statement to identify the methodologies adopted in the collection of data and the reasoning for establishing the measurement parameters and variables (location, timing, area measured, etc.). These may result from standard sampling protocols and working to accepted procedures, or they may be governed by practical constraints such as access agreements. The local planning authority (or others such as environmental health officers or Environment Agency representatives) may also have provided guidance during the scoping stage on the locations of the most appropriate points to be monitored. Where advice has been sought and provided, it should be used and acknowledged to avoid unnecessary confusion (assuming the data are relevant, accurate etc.). Increasingly, summaries of appropriate and standard data collection methodologies are being established (e.g. IEA, 1995) so that confusion over which methodology should be used and arguments over the validity of a particular approach are being removed. Specific methodologies for addressing environmental sampling are now readily available (Keith, 1996) so that guidance on any particular aspect of monitoring can be obtained.

The design of the data collection exercise or monitoring programme must be undertaken in a manner whereby the best results will be obtained to enable the predictive exercise to be successfully undertaken subsequently. It is important to identify the correct time to sample (this is probably most appropriate for ecological studies; Fig. 5 illustrates the

133

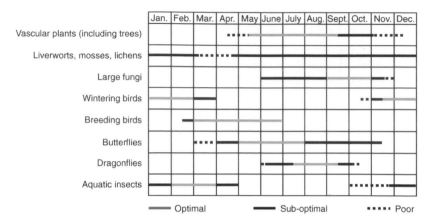

Fig. 5. Optimal ecological survey times

optimal times for ecological monitoring). Similarly, the location of monitoring points can be critical; for example, the selection of locations from which photographs of the development site are to be taken is vital for the accurate description of the landscape and visual environment, while the selection of noise-monitoring locations is important as it is necessary to identify correctly those properties that may be impacted upon by the development.

It should be standard sampling practice to ensure that the sampling equipment has been maintained correctly and methodologies used in the collection of data are valid; it is necessary to provide details in the environmental assessment of the equipment used and the dates of recent calibration. The use of equipment (such as noise monitors) that is outside its accreditation dates, is inappropriate or has not been calibrated correctly could result in the data being invalidated. The implications in terms of cost, time in collecting the data again and delays to the determination of the planning application can be severe.

Consultation

Some form of consultation is normally an essential and vital aspect of the undertaking of environmental assessments, and without the involvement of consultees the effectiveness of the assessment process and the results of the exercise can be compromised. The role and input of consultation within the environmental assessment process will vary from study to study and may involve anything from a complicated and lengthy series of meetings and presentations throughout the assessment exercise, to select meetings with the statutory bodies at the outset of the study. More and more, however, developers are realising that costly delays are inevitable and considerable public goodwill will be lost if residents are not informed of intentions well before final plans are drawn (Wiesner, 1995). Recent opinion (Wates, 1990) has been expressed that unless the role of public participation is the subject of stringent guidelines, the process will continue to be devalued and of limited importance.

Some general guidance on the Government's position on environmental issues in relation to planning can be found without consulting with the local planning authorities (through examination of the Planning Policy Guidance Notes, e.g. PPG23 Planning and Pollution Control), but these documents encourage close consultation at a local level in order to prevent unnecessary duplication and conflict of interests between the planning and pollution control authorities. Consultation has been the subject of a leaflet from the University of Manchester (University of Manchester (EIA Centre), 1995), while numerous papers have been published addressing the issue and how to approach it (e.g. Clark, 1994; Committee on the Challenges of Modern Society, 1995; Sewell and Coppock, 1976; World Bank, 1993).

There are three main groupings of consultees:

- statutory consultees that will be requested to comment formally on the planning application and the accompanying environmental statement once it has been submitted;
- non-statutory consultees who are not defined in the environmental assessment regulations but who may have a vested interest in the proposed development;
- the public (who may act either individually or as part of a non-statutory consultee).

The role of the consultee will vary depending upon the desire of the developer to include (or exclude) the organisation/individual from the environmental assessment process before the submission of the environmental statement. Once the statement has been formally submitted to the local planning authority, the developer is required to announce the planning application in a local newspaper, while the local planning authority will identify the receipt of the planning application through a variety of means but including the erection of a notice at the proposed development site. At a variety of locations surrounding the site, noticeboards announcing that an application has been received for a particular development will be placed identifying a specific location where the plans and application can be viewed. The boards will also provide a date by which comments must be received if they are to be taken into account. Such statutory notification procedures are distinct from the informal consultations that may take place during the scoping phase and environmental assessment process.

For the larger more contentious developments, awareness of the intentions of the developer may have existed for some time and it is likely that public meetings would have been held to announce the development and the company's plans. The first that any individual may know of a smaller development is when the announcement boards are erected at the site.

These represent two very different approaches and there are many degrees of consultation between them. The type and form of consultation that will be undertaken should be agreed between and within the project team at the outset of the study, and care taken not to go outside the boundaries of this agreement without the whole project team being aware of the changes that may occur.

It is generally agreed that consultation is a worthy exercise and one that will allow for a smooth-running project and reduce the potential for objections. Scoping is now accepted as 'good practice' and early consultations are considered by some to be essential (Environment Agency, 1996).

The environmental assessment regulations identify the statutory consultees. It is often appropriate to include other bodies depending upon the type of development and the issues of potential concern. For example, if nature conservation is a concern, it may be relevant to contact one, several or all of the following: Royal Society for the Protection of Birds, the Woodland Trust, the Wildfowl and Wetland Trust and the Royal Society for Nature Conservation (as well as any local ecological wildlife and nature conservation trusts). Similarly, where land use may be an issue, consultation with the following non-statutory consultees may be an appropriate and useful exercise: the National Farmers' Union, the Council for the Protection of Rural England, the National Parks Authority, the Coal Authority, the Ordnance Survey and British Geological Society.

The role of statutory consultees in environmental assessment is three-fold; firstly, they have a duty to provide to the developer, on request, any relevant information in their possession which may assist in the preparation of the environmental statement; secondly, they must ensure that the public are informed of the submission of an ES and be given the opportunity to comment on its content; and thirdly, they may advise the developer or their consultant on whether an EA is necessary for a specific project.

136

The invitation to the public to offer comments on a proposed development must be undertaken in a well-managed way to avoid public relations problems and adverse knee-jerk reactions. The Government's guidance on environmental assessment recognises that the public's concern over a development is often expressed as a concern over the possibility of unknown or unforeseen effects. The provision of information at key periods can assist in allaying such fears. The public often respond very favourably towards a development given the opportunity to comment on a proposal before it is submitted as a planning application, and indeed can often improve the route through the determination of the application. It is regrettable that many developers view the environmental assessment process as a public relations exercise; such an approach is invariably successful only if that company is genuinely committed to the undertaking of the assessment and the incorporation of comments and suggestions where they are relevant and where it is appropriate to include them.

Some developers may have concerns over confidentiality throughout the project (some of which may be due to commercial issues, others to the potential opposition to the development). They may be particularly sensitive over consultation at the early stages of project formulation and design and may have genuine concerns about placing information in the public domain.

Where the developer is a public body, similar problems over confidentiality normally do not exist and consequently, open discussion can take place during the scoping (and subsequent) stage(s).

The opportunity for the public to influence the variables of a development once the application has been submitted are minimal (unless opposition is such that the statutory consultees recommend refusal of the application until further impact mitigation is incorporated). If the public are to be involved, it should be at the initial stages of the environmental assessment.

The advantages of early consultation with the public are that the proponent can accurately inform the community of the details of the project, listen to their concerns and apply suitable weight to them in the environmental assessment. Generally, consultation before completion and submission of the environmental statement is a very worthwhile activity and positive points will almost always result. Discussions with consultees will enable access to the expertise and knowledge of certain local conditions that otherwise may not be identified.

While consultation with the statutory consultees is always encouraged throughout the environmental assessment exercise, and indeed represents good practice (DoE, 1995), it is most common for contact to be made during the scoping stage and then again before submission of the formal statement. Many local planning authorities will welcome the opportunity to review an advanced draft of the document as this can aid the identification of issues not addressed to their satisfaction or where further clarification on a specific issue of the assessment is desirable. Such an approach, while appearing to delay the submission initially, can save considerable time later if additional data collection is necessary.

Depending upon the nature of the development, confrontation with and opposition from the public may be unavoidable regardless of the real impacts that will occur from the development. The NIMBY (Not in My Back Yard) and indeed the NOTE (Not Over There Either) syndromes act as powerful public motivators to certain types of 'dirty development', such as landfills, incinerators and most types of heavy industry. In such cases it is even more important that effective and accurate scoping is undertaken so that the environmental statement cannot be faulted through the use of incorrect assessment techniques or a lack of knowledge of the site, which could be cited as reasons why the environmental assessment is inadequate.

For large developments, the establishment of working parties or discussion groups (perhaps including representatives from some of the key statutory consultees) that sit periodically during the preparation of the environmental assessment can be productive. Such working parties will enable the finished product to be an accurate and correctly balanced assessment of the probable impacts, with each of the key impacts assessed in a coherent and managed manner. In addition, these forums could be used to discuss a range of mitigation measures that may be appropriate for the site.

With large developments, the management of the consultation exercise can be extremely time-consuming and can represent a major undertaking. It is therefore important that if such a route is selected, the statutory consultees (and others who are invited to attend) must be able to see that the proponent is committed to the exercise and to the reduction of environmental impacts where it is necessary, practicable and achievable. In the event of advice and guidance being offered, it is important that such advice is acknowledged and acted upon rather than being ignored.

Some circumstances may occur when consultation cannot be carried out either as a result of secrecy requirements by the proponent (in the case, for example, of security-related developments by the Ministry of Defence) or due to commercial confidentiality issues. Where possible, such circumstances should be avoided and discouraged because of the problems of effectively and accurately scoping the study and the appropriateness of the intended development at the site being considered.

Where consultation is not possible, great care should be taken in determining the scope of the work. It is likely that the scope of the assessment will be greater than would otherwise be required in such circumstances in order to ensure that the requirements of the statutory consultees will be met at the first attempt.

However, consultation with statutory authorities can be undertaken on a confidential basis where necessary, and should be encouraged where possible (with the appropriate reassurances on confidentiality), regardless of how limited the information that can be passed from the proponent may be. Critical and fundamental opposition to a study can be identified at an early stage, significantly saving the resources and time of all concerned.

Impact prediction/measurement

There are numerous assessment techniques that can be adopted when undertaking environmental assessments; although there are a number of recommended approaches that currently represent industry best practice, there is no legislative requirement for the assessment of an environmental impact to be undertaken in any particular manner or using a specific approach. Where possible however, it is recommended that established models and methods of measurement and calculation should be used for impact prediction. It is beyond the scope of this book to undertake a review of all of the techniques that are available that could be adopted in each of the subject areas but numerous texts are available that can give guidance on this issue. There are available both general guidance (Erickson, 1994) on various aspects of impact prediction and specific guidance (e.g. Lee and Lewis, 1991; DoT, 1995) for certain types of project, identifying the issues that are likely to be of concern with that type of development, how the impacts should be assessed and what mitigation measures can be used to achieve what level of success. Such guides are invaluable for these particular types of development and contain much consideration of numerous and relevant issues that go beyond the scope of this book.

The aim of the impact assessment will be to predict, as accurately as possible (within the constraints that exist, such as time and resources), the likely environmental effects of a development. Impact prediction should go beyond comparison with national standards and should, where possible, demonstrate suitability of the development for the local situation. The selection of assessment or prediction methodologies to be used as part of the assessment will therefore depend upon:

- the issue being addressed,
- the sensitivity of the development and the significance of any impact that is likely to occur,
- the accuracy and level of detail required in the impact assessment,
- the reliability and reproducibility of the predictions,
- the availability of data that can be used in the assessment (or the ease with which they can be collected), and
- availability of time to undertake the assessment.

Impact assessment techniques can be divided into two main types: qualitative, where the impacts are largely described on the basis of previous experience (either of similar developments or in terms of the area where the proposed development is being planned) and knowledge of the likely effects; or quantitative, where an attempt is made to measure more accurately the level of impact through a numerical modelling exercise.

For example, it may not be necessary to undertake a detailed noise impact assessment for the operational stage of a development that is known to have no noise generation capability and will not create knock-on or causative secondary impacts (such as that of vehicular traffic visiting the development). Conversely, the impacts on air quality from the emissions from a proposed incinerator will require careful and detailed modelling because of the concern over the process itself (which may be emotive due to a lack of knowledge and the perception of potential harm from the facility), and also to demonstrate that the impact on air quality at ground level will not be such that health impacts may also occur and to ensure compliance with legislative limits.

All predictive methodologies are just that; they represent a *prediction* of the impacts that *may* occur as a result of the development. As a consequence, their accuracy and validity may be questioned (this in itself represents a strong reason for agreeing the methodologies to be used before the studies are undertaken). It is extremely unlikely that the precise level of impact can be predicted for any type of development, because of the very large variations and the number of assumptions that are required during the predictive exercise. For example, it is possible to predict that water quality will deteriorate if an untreated effluent is discharged to a river and this will almost certainly be confirmed upon implementation (if permitted). However, the modelling exercise that quantifies the impact is only accurate within certain statistical boundaries and it is subject to the limitations of the model used. The more precise the information required, the more difficult it is to obtain accurate information. The aim of impact prediction should therefore be to remove as much uncertainty as possible from the models used and the techniques adopted to satisfy the requirements of the study.

The use of mathematical models to predict an impact is subject to potential errors at all stages of their development, calibration, validation and use, because of the wide variability in the environment being predicted (for example, air quality will vary by the minute) and the inability of the models to reflect accurately the huge variations that can occur. However, improvements in model preparation and expertise of those using them can enable valuable predictions to be undertaken; in the absence of alternative methodologies, choices are limited. Publications providing an introduction to all aspects of environmental modelling are now available (Schnoor, 1996).

Once impact prediction has been undertaken (either qualitatively or quantitatively), it is necessary to consider the issue of significance: how significant will that impact be on the sensitive receiver? What is the relative scale of impact? Significance is a site-specific consideration that will experience huge variation; an odour that is considered as acceptable and the norm in a Third World city (for example, arising from a failure or non existance of the sewerage system) is likely to be unacceptable in a residential area in Surrey.

In addition to addressing significance in terms of the existing and surrounding environment (data on which will have been collected previously), it is possible to compare the likely impacts with a wide variety of standards and limits. A whole host of environmental legislation in the UK in the form of regulations provides limits which must be complied with. In addition, many standards have been proposed and adopted by a wide variety of organisations (such as the World Health Organization) that have been given 'best practice' status by determining authorities and statutory consultees. A review of relevant standards has been prepared (Bourdillon, 1995).

The following sections consider and review some of the techniques and practices involved in prediction of environmental impacts.

The water environment

Many different models are available for assessing impacts on the water environment; these include the assessment of impacts on water quality as well as on flow characteristics.

Impacts on water resources is one area where a host of simplified and complex models are available; these include rainfall run-off and catchment models where the impacts of flooding as a result of continued urbanisation, for example, can be predicted. The impact of an effluent released into

141

a water body or aquifer can be modelled and can result in the identification of the dilution capacity that would be required before the body of water would return to normal (and any state up to that). Such a model may also provide results that represent a useful indicator of impacts on the ecology of the water body (from experience, appropriate experts would be able to identify which species would be able to tolerate a certain level of pollution).

Prediction of flood risk and run-off characteristics, if identified as a potential issue of concern as a result of a proposed development, invariably require the use of a model to simulate the ambient conditions and parameters which then permit the addition of model characteristics so that the impact can be predicted. The success of the model in such an exercise is dependent upon a number of factors — predominantly:

- estimation of parameter interactions, including consideration of all processes and conditions within the study area;
- the availability of accurate data;
- the specificity of the model itself (models developed for one particular site or specific purpose are difficult to adapt to other circumstances).

Surcharging of existing drainage facilities is also a relevant issue that is required to be addressed where that discharge is taking place to a foul sewer or surface drain. The capacity of existing facilities to receive new discharges must be assessed, and the ability of downstream treatment systems to treat the additional effluent (in the case of effluent discharges) also require consideration. Unless the development is likely to produce a large volume of high-strength effluent, it is likely that the capacity of the receiving facilities will be sufficient. However, this must be confirmed with the sewage undertaker who will advise of negative impacts that may occur and the suitability for the discharge to occur. It will be necessary to estimate the quantity, strength and flow profile of the effluents to be disposed of and compare these data with the system design capacity and existing flows.

The accuracy of hydrological models has been demonstrated to decrease as their size increases (Wood *et al.*, 1990), as can be expected because of the probable increase in the complexity of interrelationships of climate, vegetation, soil conditions and topography of larger catchments. However, modelling of catchments as large and complex as the Severn and Thames has been attempted with some success (Jolley and Wheater, 1996).

Impacts on water quality from a development will obviously depend upon the type of development and the effluents or wastes that may be produced (and the stored materials that may be spilled or accidentally released), but they will also be dependent upon the type and quality of the receiving water body.

The types of impact that will require assessment include those from both point sources and those from diffuse sources; the latter are more difficult to predict as part of the assessment process and there are fewer models available for predicting the impacts that may occur. Simple calculations can be undertaken by a variety of methods, including the unit-flow method and the concentration versus flow methodology. More complex models are available that enable a prediction of changes in water quality, but the complexity of hydrological environments often results in inaccuracies and a lack of confidence in the results obtained.

For smaller developments where only minimal impacts on water quality are likely to occur, impacts of known or scheduled releases can be assessed without complex modelling and can be gauged for their satisfactory inclusion within the environmental statement through the reaction of the regulating authorities; all discharges to controlled water bodies (defined in EPA 1990 as any inland water body and coastal water) will be subject to the approval by the Environment Agency for discharges to water bodies and by the local sewage undertaker if the disposal is to the sewerage system.

Licences enabling discharges to take place will be awarded only where the effluents fall within certain guidelines; in the event of the discharge being approved, it can be assumed that it will have only a minimal impact and will not significantly increase the subsequent impact following effluent treatment. For example, the discharge of a small volume of low-strength effluent into an existing sewerage system cannot be said to have zero impact but it will not have a noticeable impact upon an existing treatment system or on the receiving water body (following passage through the treatment process) because of the flow that is already occurring. A worsening and measurable impact on the ultimate receiving water body will only occur if the volume of new flow or its pollutant content is significantly different from that already being received from other multiple sources. The licensing authorities will however require information not only on the nature of the predicted and normal discharges, but on what other discharges could take place as a result of plant failure or accidental releases and what control measures have been included within the system or facility design to guard against such occurrences.

Mathematical models of groundwater systems can be useful tools that can assist in the assessment and prediction of environmental impacts. Models vary in complexity from simple equations which have a direct solution (termed analytical models), to sophisticated finite element or finite difference models (termed numerical models) which use iterative techniques to obtain approximate solutions to the governing differential equations.

The most basic groundwater models are designed to represent groundwater levels and flows under steady-state conditions in one, two or three dimensions. More complex models may also simulate time-varying effects and additional processes such as particle movement, solute transport, groundwater chemistry or density, flow in fractures etc.

Groundwater modelling studies may be divided into several stages:

- Step 1: Develop a conceptual hydrogeological model.
- Step 2: Select appropriate modelling software.
- Step 3: Set up an initial model using site data where possible.
- Step 4: Calibrate the model against observed data.
- Step 5: Carry out model runs to test the sensitivity of the model to various parameters.
- Step 6: Use the calibrated model to carry out predictive simulations as required.

Depending on the complexity of the project, each stage may take only a few hours or it could take several days or months. These stages are discussed in more detail below.

The first stage in a modelling study is to collate all available data and develop a 'conceptual' hydrogeological model of the site. The conceptual model would identify the aquifer(s) beneath the site and would assess the principal components of a water balance for the site and surrounding area. The extent of area which should be considered depends on the hydrogeological setting and on the objectives of the project. In general, the area being considered should be at least large enough for activities or developments at the site to be unlikely to have any significant effect at the boundary.

Having developed a conceptual understanding of the hydrogeology of the site, an appropriate model is then selected, depending on several considerations including:

- The amount and quality of available data. It would be a mistake to use a complex model for a situation in which the available data, with which to set up and calibrate the model, are sparse or unreliable.
- The objectives of the assessment, and the extent to which modelling could assist in achieving these objectives.
- The available resources in terms of time, budget, technical skills and experience and computing hardware and software.

An initial site-specific model is then developed, using site data where possible. The selection of appropriate boundary conditions is particularly important and should only be undertaken by an experienced hydrogeologist. There will normally be some parameters — such as hydraulic conductivity — whose values are not fully known, so initial values for these should be estimated based on literature values or local experience. The model is then 'run' using the initial set-up, and the results compared against observed data such as groundwater levels, river flows or contaminant concentrations. The values of the unknown parameters are then adjusted, using an iterative approach (known as model calibration), until an acceptable match is obtained between observed and simulated conditions. An example of the output of a calibrated groundwater model, recently completed by PB Kennedy and Donkin Ltd on behalf of the UK Environment Agency, is illustrated in Fig. 6.

There is never sufficient information or data for perfect calibration of a model, so some uncertainty concerning parameter values will always remain. The model is always an approximation and simplification of reality. It is therefore advisable to carry out a sensitivity analysis on the calibrated model, to test the dependence of the results on the values used. In this way some estimate of the possible precision of the results can be obtained.

It is only following all these stages that the model can be used to predict what may happen in the future, given various 'scenarios' in terms of site development, etc. The results of a 'no action' run can be compared with the results of alternative development options to assess the effects or impacts of the options.

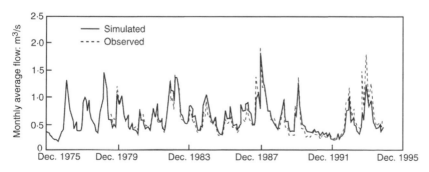

Fig. 6. Observed and simulated river flows

Air quality

Assessment methodologies for determining potential air-quality impacts can be either quantitative (through the use of a computer model that attempts to predict a concentration of a pollutant at a given point), or qualitative (where approximations or estimates of likely impact are attempted).

The latter is more common in those situations where impacts on air quality are not considered to represent a likely or significant impact and where a review of the activities and potential pollutants involved will be satisfactory. This is often appropriate in the estimation of dusts from construction activities. Although computer models are available for modelling this type of activity (e.g. the fugitive dust model developed by the United States Environmental Protection Agency (USEPA) which calculates dust concentrations based on emission factors for each type of activity, dust type and wind speed), unless the site is near to a particularly sensitive location (e.g. schools or hospitals) or has a particularly large number of construction activities taking place, such an approach is often unjustified.

In such instances, the clear identification of potential dust sources that may represent a source of dust and the incorporation of mitigation measures to reduce the potential for dust nuisance will be sufficient. This approach is recommended to estimate dust impacts from surface mineral extraction sites (DoE, 1995c).

Modelling of air-quality impacts enables plant designers and planners to examine the predicted performance of a proposed development, to assess the effectiveness of mitigation measures and to make predictions concerning future air quality. Air Quality Management is becoming an important tool in the regulation of new developments; proposed emissions are not considered in isolation but in conjunction with existing data, modelling of emissions and the development and maintenance of emissions databases. Some of the models available and the constraints on their use have been reviewed (e.g. Gould, 1996), but all will have their advantages and disadvantages and will represent approximations of reality.

The Gaussian model is the most common of the numerical dispersion models used in environmental assessment and can be used for predicting the concentration of pollutants away from the generating source. The model takes into account how an emission of a known concentration is translated to a ground-level concentration as a result of dispersion by wind and air currants and by dilution.

A number of runs of the model are usually required to predict the

146

short-term pollution impact (manifested as acute pollution impacts) as well long-term impacts (indicative of chronic situations). The former enables prediction of worst-case hourly average concentrations whereas the latter may be used for prediction with annual average concentrations; both results can then be compared against the appropriate air-quality standards in order to identify points that may be exposed to concentrations of pollutants above those considered acceptable. The maximum ground-level concentration of pollution released will occur at the point where the plume first grounds.

The USEPA has developed the UNAMAP suite of models, which are in common use in the UK and elsewhere. These include short- and long-term variations (ISCST and ISCLT respectively), which enable them to be used for a variety of applications. Recent versions have included features that can take account of momentum-induced plume rise and buoyancy plume rise, calculated as a function of distance downwind and building wake impacts (which can both affect emission dispersion).

For the modelling of point sources of pollutants it is necessary for the following information to be known:

- the temperature of gases at the point of exit;
- the stack height;
- the internal diameter of the stack/chimney;
- the exit velocity (or e-flux velocity);
- the concentration of pollutants;
- the presence of nearby buildings and their height; and
- the surrounding topography.

In addition, details of the atmospheric conditions are required to identify prevailing weather conditions and, atmospheric stability (or the amount of turbulence and air movement). Generally, stable conditions and low wind speeds result in poor levels of plume dispersion. With this information, a worst-case scenario can be modelled that will predict the concentration of pollutants at required assessment points. A typical print-out from such a model is illustrated in Fig. 7. Print-outs are normally produced at the same scale as an Ordnance Survey map and the contours can be printed directly onto a map or subsequently overlain.

The reliability of air-dispersion models as a method of predicting ground-level concentrations has, recently been called into doubt in a recent study by the Environment Agency, however (DoE, 1996). The study has identified the weakness of many models for predicting short-term concentrations, particularly where the plume is grounding. Other models

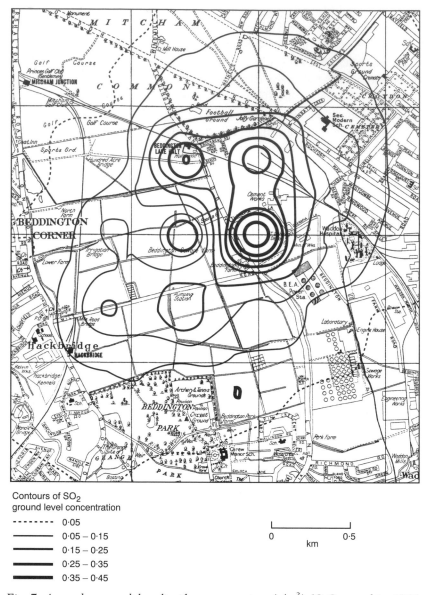

Contours of SO$_2$
ground level concentration

- - - - - - -	0·05
———	0·05 – 0·15
———	0·15 – 0·25
———	0·25 – 0·35
———	0·35 – 0·45

Fig. 7. Annual mean sulphur dioxide concentrations (g/m³) (© Geographia, 1964. Reproduced by permission of HarperCollins Publishers Ltd)

(such as the UK–ADMS model developed by Cambridge Environmental Resource Consultants) have been developed that achieve a more realistic calculation of plume use and dispersion which results in more accurate predictions of pollutant concentrations downwind of a stack. Such models are currently receiving considerable recognition from the regulatory bodies and local authorities.

In addition to those for point-source models, predictive models are also available for area sources and for line sources. The former can be used for predicting dust problems from construction sites and will consider sources such as dust pick-up from vehicles operating on the site as well as fugitive emissions from stockpiles and other dust-producing activities such as excavations and earth handling. Line-source models are used for predicting the likely concentration of vehicle emissions from a proposed new road in existing residential areas. Dispersion models such as CALINE4 (developed by the Californian DoT), or PAL (a screening model for point, area and line sources) operate with key data such as traffic flow and composition (mean vehicle speed, flow rate and type, i.e. percentage of cars and HGVs) and assume various meteorological conditions. A variety of different pollutants can be modelled: predictions can be made of the concentrations of pollutants at key points and estimates can be given of the number of days in any year when relevant standards may not be complied with.

The majority of the Gaussian models operate as flat-ground models, i.e. they cannot take account of undulating or sloping topography. Such features could be key elements in predicting the impact of, for example, an emission from a factory located on a valley floor on houses on the adjacent valley sides. Refinement of the models to take into account such specific situations is possible and there will be greater confidence in the results obtained.

The more complex the models, however, the greater the is potential for error and inconsistencies to occur, and the less accurate the predictions may be. Similar inaccuracies will occur for the area- or line-type models. It should be noted that these models represent approximations of reality and may not take into account the numerous atmospheric interactions that are likely to occur. The limitations of the models should be understood and made clear in the results of the environmental assessment.

Ecology

Guidance has been published (Beanlands and Duinker, 1983b; English Nature, 1994, 1995) that identifies good practice in the assessment of ecological impacts from a development. Although they do not refer to specific methods of data collection or predict the impacts that may occur, this guidance does provide essential and valuable advice on the protocol of ecological assessment. A selection of the processes involved are reproduced in Table 12; although some of these appear obvious they are often overlooked in the assessment process.

The predictive methodologies for identifying the impacts on the ecology of a site will be largely dependent upon the site itself and the nature of the development; a rural site (not covered by agricultural land) will require a greater complexity of study than a derelict industrial site because of the potential for locally, regionally, nationally or internationally important species to be present (derelict sites must be considered with especial care because of the potential for rare species to be present due to the lack of public access to such sites). The aims of the predictive exercise are two-fold: firstly, to quantify all the impacts that are likely to be associated with the proposed development; and secondly, to identify as closely as possible the effects that will occur on the ecology.

Prediction of impacts on ecology of a site can be undertaken only when it is clear what species/habitats are present. This does not necessarily indicate that a full ecological survey is required for every site; guidance on the type of habitats at a site, the species likely to be present and the importance of each can be provided by English Nature (for example), local ecological trusts and local ecological consultants. Once the likely importance of a site has been gauged, the appropriate level of survey can be undertaken. This can vary from a Phase 1 habitat survey where a general description of the habitats present is prepared (referring to species composition) to a Phase 2 survey where the importance of species present is assessed and a comparative evaluation of habitat types is undertaken. The most detailed survey (Phase 3) requires intensive sampling to identify and quantify species populations. Invariably Phase 3 is undertaken on specific areas of a larger site in the event that earlier studies have identified the need.

Impact prediction on ecological resources relies mostly on expert opinion and the specific knowledge of specialists. It should also be noted that it is unlikely that one individual will be able to offer sufficient expertise on all matters at a complex site. The IEA states that 'it is of paramount importance that individuals only undertake survey work within their area of competence' (IEA, 1995a). It is also important that the each of the ecologists does not work in isolation from the other team members, as there is significant overlap with many other disciplines within the environmental assessment.

A number of techniques can be used during the assessment process to predict impacts; these will include matrices, checklists and flow charts, which are most useful either during the scoping stage or during the predictive stage to ensure that all interactions have been considered.

Table 12. Summary of good practice in environmental assessment with regard to nature conservation

Area	Examples of good practice
Initiating the assessment work	Establish effective communication within design team. Early consultation with English Nature (EN). Avoid areas of nature conservation value.
Determining the scope of the assessment	Agree baseline requirements. Undertake scoping exercise in a systematic manner. Undertake consultation with EN.
Characterisation of the baseline environment	Ensure team is suitably qualified. Ensure level of detail is sufficient. Ensure satisfactory timing of studies.
Project description	Give clear description with maps, plans etc.
Impact prediction	Duration, timing, probability, reversibility and susceptibility of mitigation along with any cumulative impacts that may occur (due to other development) should be assessed. Specify short-, medium- and long-term impacts. Elements of wildlife, their importance and sensitivity, and ability to escape should be considered.
Assessing the significance of impacts	Evaluate as a function of scale, severity, value and sensitivity. Consider scientific importance, amenity value and rarity (locally). Explain how and why impacts are significant.
Mitigation of the impacts of the development	Preserve nature conservation where possible. Ensure measures do not have implications for other conservation resources. Be clear on commitments regarding after-uses. Enhancement or replacement schemes should be appropriate.
Presentation of information in an environmental statement	Make clear and concise. Make primary ecological data available.

The significance of ecological impacts should be described in terms of the sensitivity of the impacts, the value of the ecological systems affected and the magnitude of those impacts (English Nature, 1994).

The importance of a species is normally described in terms of its significance in the local community, i.e. is it of local, national or international importance? For example, a species that is rare in the local environment but common nationally will be of less significance than a species common locally but rare internationally. Other equally important aspects to consider are whether species that are present at a site are protected by legislation and are therefore required to be assessed in such terms (for example, it is an offence under the Protection of Badgers Act 1992 wilfully to kill, injure or take a badger or attempt to do so, and sett disturbance can only be carried out under licence (Cox, 1993). The legal protection of a species or habitat is no guarantee of their protection, however; a recent housing development in Peterborough has resulted in the relocation of 15 000 great crested newts (at a cost of approximately £1000 per newt). The development has proceeded despite the existence in the UK of the strictest current EU nature protection laws (*Daily Telegraph*, 1996).

Ecological impact assessment should not focus only on the value of species themselves but also on the habitats and ecosystems of which they are part. The sensitivity of an ecosystem should be considered; the local species may not be considered rare either locally or internationally, but impacts from the development may result in destruction of the habitat as a result of, for example, installed drainage systems. The value of species diversity is also an important aspect to consider; an ecological survey for a site in Cornwall recognised that a portion of a site was of great importance, not because of the presence of a rare species but because of the mosaic of habitats that was present in close proximity to each other (ECC International and Rust Environmental, 1995).

Other criteria that may need to be taken into account depending upon the site and its characteristics are:

- ancient woodlands or habitats with historical importance;
- sites of cultural importance;
- sites with amenity value (that are not designated landscapes, for example);
- areas with aesthetic appeal; and
- sites that may have potential for improvement.

A brief evaluation of the methods of impact significance and assessment for flora and fauna has been published (IEA, 1995a) and this document is invaluable to both the non-specialist looking for guidance on the procedures to be adopted, and to the specialist for the promotion of best practice.

Noise

Assessment of noise impacts and the significance of any impact as a result of a development is dependent upon a number of factors such as the ambient or background noise levels in the vicinity of the site, the type of development and its operating characteristics. The aim of noise impact prediction is therefore to identify the ambient conditions, predict the increase in noise levels (or change in characteristics) as a result of the development and assess the significance of that impact.

Noise impacts will generally occur from three sources, depending upon the nature of the development being assessed:

- noise from static plant such as generators or air-conditioning units,
- mobile noise generated from plant at a site as a result of a specific activity such as loading of glass cullet and
- traffic noise on roads or at sites where deliveries (for example) are taking place.

Assessment of noise impacts will require each of these noise sources to be identified and assessed where relevant as part of either the construction, operational or closure/restoration phases. This is necessary as there are likely to be different components contributing to the noise environment during each stage, and the impact of each is more likely than not to be of differing intensity and significance.

Collection of baseline (ambient) noise data is the first element of the predictive exercise; without such information it is difficult to assess whether the new development will result in an increase in ambient noise levels. The same development may cause a significant impact in one area where noise levels are relatively low, whereas in another, more developed area, the same development may have noise levels lower than the ambient conditions as a result of other noisier sources being present and more dominant; on noise grounds alone, the latter site would be a more appropriate location (assuming, of course, that the total noise level is not already too high).

In the event that noise data are not available and for some reason they cannot be collected, it is possible to predict the impact based on professional experience and a thorough knowledge of the site area. In the absence of any sensitive receivers, such as residential properties, schools, churches and hospitals, it is likely that a higher noise level would be permitted than in an area where any (or all) of the above are present in close proximity. Alternatively, if a factory unit is proposed for location near a residential area that does not currently have any other similar industries or any road networks, it is likely that the development would result in an unacceptable rise in ambient conditions (it should be noted that in such a circumstance noise would probably be identified as one if not the most important issue and it would be necessary, if not essential, for ambient noise levels to be collected and a detailed assessment undertaken if the planning application were to be given any serious consideration).

Prediction of noise levels arising from a development (in either the construction or operational phases) is a relatively straightforward and methodical process whereby each of the sources of the noise are identified, sound power levels for those activities, processes or plant are determined and elements that may affect the final noise level (such as enclosures, barriers, etc.) are taken into account.

There are a variety of sources of noise-level data; equipment suppliers quote sound power levels of their equipment (such as compressors, or on a larger scale, back-actor excavators), while various publications provide considerable detail on the noise levels from particular pieces of plant or from a particular activity (BS5228 gives details of sound power levels for several hundred items of construction-related plant). The IEA and the Institute of Acoustics are preparing a document addressing the methodologies for noise impact assessment as part of environmental assessment studies; it is hoped that this document will provide clear guidance on the steps to be followed in conducting such an exercise (Turner, 1996).

Another methodology for collecting data to enable prediction of noise impact is the measurement of the activity in question at a site where a similar activity is already operating. Care should be taken during such activities to ensure that accurate measurements are collected (i.e. that the result is not influenced by other activities or nearby plant) and that they accurately reflect the proposed activity that is being assessed. In the instance where data for an activity does not exist, and measurement of an existing operation is planned, it is often prudent to invite a representative of the local planning authority (LPA) or the environmental health office to attend to validate the veracity and relevance of the results collected.

Once the noise levels for an activity at a development site have been predicted, the impact on sensitive receivers can be gauged. It will mainly be a function of the distance to the sensitive receivers and depend also on whether there are any structures that will act as noise barriers (such as topographical features or other buildings) although other elements can also affect the extent of impact (such as ground absorption and vegetation).

Distance attenuation is often the most critical element in achieving noise attenuation and can be calculated easily if the source noise level and the distance between the source and receiver are known. Several documents provide guidance on the calculation of distance attenuation factors (for example BS5228) but Table 13 gives an indication of the correction factors that can be expected for a variety of distances from the source.

Placement of barriers can provide a shield to sources of noise, resulting in a reduction at the receiving point. MPG11 (DoE, 1993) suggests a reduction of 5 dB(A) where partial shielding occurs and a reduction of

Table 13. Approximate correction factors for given distances from a noise source

Distance (m)	Correction (dB(A))	Distance (m)	Correction (dB(A))
1	8	24–26	36
2	14	27–29	37
3	18	30–33	38
4	20	34–37	39
5	22	38–41	40
6	24	42–47	41
7	25	48–52	42
8	26	53–59	43
9	27	60–66	44
10	28	67–74	45
11	29	75–83	46
12	30	84–93	47
13	30	94–105	48
14	31	106–118	49
15–16	32	119–132	50
17–18	33	133–148	51
19–21	34	149–166	52
22–23	35	167–187	53
		188–210	54

10 dB(A) if the source is completely screened, but these are rule-of-thumb guidelines and the actual reduction will depend upon the density of the barrier, its effective height in relation to the source and receiver, its length and its proximity to the source (the closer the barrier is to the source, the more effective it is likely to be). The presence of tree barriers will be of limited significance in reducing noise (beyond the psychological effect they may have), and it is recognised that a screen of trees would need to be over 30 m wide for there to be any significant attenuation effect (BS8233,1987). This aspect is addressed further under 'Mitigation measures', below.

Ground absorption correction factors are a relevant consideration where the ground is predominantly soft; for example, a reduction of 8 dB(A) may be relevant for a sensitive receiver located some 250 m away across fields.

During the impact assessment it is necessary to predict the levels of noise from the development at all stages of its operation. This includes the construction phase, where typically large numbers of noisy plant are operating in an intensive manner for short periods of time, and the operation phase, which may produce a lower noise level but over a longer period. Operational activities could also create an increase in noise levels during previously quieter periods (for example, the use of a site as a milk distribution yard is not likely to increase daytime ambient noise levels but will result in additional noise levels between 4.00 and 6.30 a.m. as the floats are loaded and they leave the yard).

Impacts during the closure or restoration phase are particularly relevant for developments such as landfills. Their impact can be pronounced as they may represent the noisiest activity in terms of maximum noise levels: noise attenuation barriers are being removed, and plant are operating on site boundaries. On the positive side, however, the activities will be of a short-term nature and will represent the completion of the site.

It is also necessary when assessing noise impacts to consider the noise profile, in terms of both its periodicity and its frequency. Noise from heavy plant can often be reduced by the use of silencers, although their efficiency for decreasing lower-frequency noise is more limited. Similarly, short high-pitched noise can often be more disruptive (such as steam release chimneys which 'whistle', or reversing bleepers fitted on HGVs).

Prediction of traffic noise impacts (as a result of new road sections or of increases in traffic flow associated with a new development) is often an essential component of the assessment. The statutory approach to this form of noise impact assessment is available in the UK (Department of Transport, 1988) but guidance on the significance of the results obtained is not available for properties affected by increased traffic flows on existing roads. Impacts of new roads are often considered in terms of the number of properties affected and the degree to which protection (in the form of double glazing, etc.) may be required. For existing road developments a doubling of the traffic flow will generally cause an increase of 3 dB (subject to the percentage of HGVs present).

The assessment of significance of an identified noise impact is generally a site-specific issue, but one to which local authorities are increasingly applying their own developmental control standards. BS4142 states that an increase of less than 3 dB is not likely to be noticeable, one of 5 dB is identifiable but likely to be of marginal significance, and an increase of 5–10 dB over the background is likely to result in complaints. The focus of most noise impact assessments will therefore be to minimise the noise impact at the sensitive receivers to an increase of below 10 dB. Where short-term activities are taking place (typically during a construction programme), higher levels of noise will be permitted, as referred to in MPG11 (DoE, 1993a) particularly if the activity will enable lower noise levels during the subsequent periods of operation (for example, the construction of a perimeter bund around a mineral extraction facility). In such cases it is common for mitigation measures to be adopted as part of the development, to reduce the scale of impact; these are subsequently reviewed.

Landscape and visual assessment

The subjectivity of this issue and the difficulty in quantifying this element of environmental assessment invariably results in it being given scant attention in environmental statements. However it is probably one of the most important aspects, as a development will often result in a physical and tangible addition to the landscape. The concern over the lack of objectivity in some of the earlier assessment techniques has resulted in the development of methodologies that enable a clearer examination and assessment of the possible impacts.

To undertake an assessment of the likely impacts from a development, it will be necessary to prepare a number of visual aids that represent the site and its surroundings. A variety of methodologies can be utilised either singly or in continuation. They include aerial photos of the site, photographs and video images into and from the site, plans and maps illustrating important features such as footpaths, viewpoints and areas of designated or protected landscapes. These enable an accurate representation of the existing environment to be made, as well as being useful for the subsequent impact assessment stage where an image of the development can be superimposed upon the 'before' image. One of the most basic and frequently used devices (because of its effectiveness in identifying geographical impact extent) is the zone of visual influence (ZVI). The use of such figures can quickly identify and present the areas from which a development will be visible. An example of a ZVI for an aggregate extraction proposal is illustrated in Fig. 8.

It will be necessary in the impact prediction stage to describe or illustrate the development in terms of its magnitude, significance and likelihood of occurring. There are a number of standard approaches to such assessment (IEA and LI IEA, 1995; Countryside Commission, 1991) and the use of such standard and accepted approaches is recommended unless considered inappropriate for the development in question or an improved approach has been developed for the particular development type.

The magnitude of the impact from the development will be a function of its physical size and location within the environment, the materials used (particularly their colour, texture and appearance) and the time scale for construction. Although these elements can be described, the most useful way to present them is through a visual methodology, which can range from simple drawing elevations and layout plans (prepared as part of the design of the plant) to cross-sections through the intended development including surrounding land forms, to artist's impressions of the development, to photomontages of the development from a number of key points and (more recently) to simulation of the visual environment in virtual reality. These techniques are mentioned here in ranked order of increasing complexity (and cost) but also in terms of their increased ability to illustrate the development accurately. Examples of the more advanced techniques are presented in Figs 9 and 10.

The construction of models for use at public meetings and exhibitions is also useful for illustrating the effects of some developments; for example, two models illustrating the before and after landscapes were constructed by ECC International to demonstrate the final landform of a china-clay tip in Cornwall, and are illustrated in Fig. 11.

The impact significance of a development can be defined simply as the impact on the visual environment as experienced by those sensitive to changes in visual amenity, i.e. the local population, walkers, tourists and other amenity users such as fishermen or picnickers, and vehicle drivers and their passengers. The impacts should be described in terms of the number of people in each group that are affected but also in terms

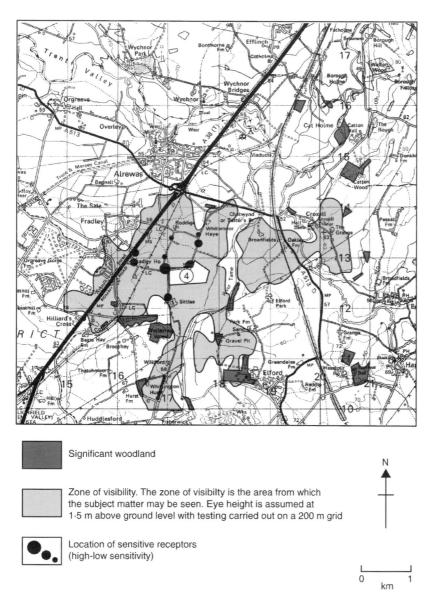

Significant woodland

Zone of visibility. The zone of visibilty is the area from which the subject matter may be seen. Eye height is assumed at 1·5 m above ground level with testing carried out on a 200 m grid

Location of sensitive receptors (high-low sensitivity)

N

0   1
km

Fig. 8. Zone of visual infuence for an aggregate extraction proposal (based upon the 1985 Ordnance Survey 1:50 000 Landranger map with permission of the Controller of Her Majesty's Stationery Office, © Crown Copyright MC 88595M0001)

Fig. 9. *Photomontage showing 'before' and 'after' appearance of a proposed development (courtesy of David Jarvis Associates)*

160

Fig. 10. Computer-generated image of a proposed commercial estate (courtesy of David Jarvis Associates)

of the type of impact: each group will experience a differing level of impact based on an individual reaction to the development but also on the level of exposure to which they are exposed. The Department of Transport (1993) prepared a scale of impact significance to enable a descriptive assessment of how much each group will be affected:

- no change — no discernible deterioration or improvement;
- slight impact — a barely noticeable deterioration or improvement;
- moderate impact — a noticeable deterioration or improvement;
- substantial adverse or beneficial impact — a significant deterioration or improvement in the existing view.

Evaluations of significance, particularly for the subjective elements of landscape and visual assessment, should be clearly defined. Government guidance states that

> The conclusions on significance . . . will not automatically be accepted . . . by the Planning Authority or general public especially where judgements about degrees of nuisance or unquantifiable aspects are concerned'

> (DoE, 1994c)

The value of the landscape must also be taken into account in any landscape assessment. The value of a particular landscape to an individual will vary considerably but can be considered to be a function of the number of different types of landscape that are present and their diversity, the presence of man-made features and the relative age of the landscape. For developments where substantial landscape impacts are likely to occur as a result of a development, it is often worthwhile for the landscape

(a)

(b)

Fig. 11. Models of a proposed china-clay tip in Cornwall (ECC International): (a) with development; (b) existing situation; (c) post-restoration (courtesy of ECC International)

(c)

assessment to be conducted in association with the public in the area of the site. This can be done in a non-specific manner using question-naires requesting general comments on the landscape, through to more direct consultation requesting the reaction of a group of people to the proposed development. The Landscape Institute, in their recommended guidelines for undertaking assessment of landscape and visual impact, state that conclusions should be partly based on views expressed during consultation exercises.

Archaeology and cultural assets

Impacts on these resources are required to be assessed in two main groups; those that are known and those that are yet to be discovered. Impacts on the latter are obviously harder to identify until they are proved to be present and the nature of the remains and their condition have been confirmed. Impacts on the former, i.e. known archaeological sites or historic buildings, can be direct (for example, obliteration of the site within the development footprint) or indirect (such as vibration impacts that may occur as a result of a new access road). Indirect impacts are often not as obvious as direct ones, and careful study of the proposals can be required to ensure that full consideration of all potential impacts has been included.

Assessment of impacts will often be required to be undertaken by a specialist; this is particularly important where archaeological remains are identified as the result of a survey, or where the potential for remains being present is high.

Impact assessment is normally undertaken on the basis of information that is already available. The most common source is the Sites and Monuments Register but sources can also include aerial photographs, historic Ordnance Survey maps and publications on the historical past of an area. The other main source of information is invasive surveys at the site itself.

Significance of impact is most often addressed in geographical terms; The Department of Transport (1993) identifies four categories of importance for archaeological remains:

- sites of national importance,
- sites of regional/county importance,
- sites of district/local importance and
- others that are too badly damaged for inclusion in the above.

In order to assess the precise significance and extent of impact, however, it will be necessary in an assessment exercise to consider the development impacts on the archaeological site in terms of its rarity, its condition and its resilience. An environmental assessment of a waste-transfer station was required to address vibration impacts from waste deliveries on a precariously balanced stone structure whose worth would have been severely reduced had it been toppled (Rust Environmental, 1995).

When sites of interest or value in the vicinity of the development site have been identified (either through their presence on the Sites and Monuments Register or through invasive surveys that identify new sites), most local authorities will require an assessment to be made of the potential for other archaeological remains to be present in the area. This would normally be discussed as part of the scoping study and would enable the invasive survey of the site (if one is deemed to be necessary) to be designed to cover the development area of the site in sufficient and satisfactory detail.

There is clear overlap between the landscape and archaeological/cultural heritage aspects of the environmental assessment where the latter are identified to be of concern. For example, the context of a National Trust property or major archaeological site could be severely impacted upon by some types of development which could have important repercussions on the viability of that resource as a tourist attraction.

Impacts on items of cultural heritage such as listed buildings, conservation areas and historic parks or gardens can often be addressed in a similar manner to those on archaeological remains, with consideration of a number of other parameters such as building age (or period), the presence of associated buildings nearby and the diversity of the historical background at the site and nearby (DoE, 1990a).

Traffic

New road developments will clearly require a detailed environmental assessment. Direct traffic impacts will represent a significant element of both the justification for the new link as well as the indirect impacts that will occur on other elements of the environment (noise, air quality, ecology, severance etc.). The DoT has issued clear and thorough guidance on the assessment methodologies to be used (Department of Transport, 1993b) and although these do not specifically refer to developments other than road networks, some elements of the predictive methodologies that are recommended can be applied to other schemes and developments.

For existing roads that will be subject to additional traffic resulting from a proposed development (and not be covered by the DoT procedures), a large number of aspects should be considered but they will be site-specific and depend upon the development, the location of the site and a variety of local circumstances. A publication by the Department of Transport (1983) suggested that an increase in traffic flows would have a varying impact; a 30% increase would result in a slight impact, a 60% increase would create a moderate impact and a 90% increase would cause a substantial impact. Increases of less than 10% have been identified as creating no discernible impact (IEA, 1993). These are very generalised assessments of impact severity that do not take account of the specific impacts or the location and type of sensitive receivers, but they are useful at the initial assessment stage for gauging the approximate or likely level of impact.

Traffic impact assessment will require an assessment of vehicle movements into and from the development, the make-up of the traffic (i.e. the percentage that are HGVs), the routes that the traffic will take and how each of these affect the local road network. Although the main focus of the traffic assessment will be to confirm whether the existing highway network has sufficient capacity to handle the additional flow, it will also be important to identify the need for any improvements that will be necessary in terms of safety, junction layout, road surface improvements and sineage.

The impact prediction exercise will be required to be undertaken at all stages of the development, since vehicle numbers or flow profiles will vary. For example, the construction stage of a development may result in large numbers of HGVs delivering raw materials and plant and removing excavated soil, whereas the operational phase may be dominated by passenger vehicles. Similarly, impacts may increase with time during the operational phase as the development expands its client base to attract greater numbers of cars. The development of an exhausted quarry into a landfill will result in continued visits to the site of HGVs, but the times of visiting and the number of visits involved may differ, as well as the duration of each visit. Separate assessments should therefore be undertaken when significant phases of the development take place.

Guidelines have been published that document the methodologies that should be used to identify how traffic from developments impact upon the road network (Institute of Highways and Transportation, 1994; IEA, 1995).

The assessment of traffic impacts should enable people who are significantly affected to identify the worst environmental impact that can be expected. The assessment should also indicate how often these conditions may occur. The Institute of Highways and Transportation (IHT) recommend two criteria that will determine whether particular sections of the highways should be subject to environmental assessment:

- if traffic flows increase by more than 30% in the opening year as a result of development traffic;
- if other sensitive areas are affected by traffic increases greater than 10%.

Assessments should be undertaken for the first full year of operation and also when additional phases result in the above limits being exceeded. The aim of the traffic assessment in the environmental assessment process should be to identify when the worst level of impact will be generated and to define normal conditions, i.e. the baseline or existing situation. The assessment of worst-case conditions should therefore be locationally defined and specific in terms of effect (IHT, 1994).

The requirements of the traffic assessment will vary depending upon the type of development; for example, the proposed location of a shopping centre will require careful and detailed study of the traffic generation rates at peak times during the working week and at weekends. A variety of databases can be used to predict generation rates and to aid the estimation of traffic flows. Similarly, the assessment should be undertaken for a date in the future by estimating the annual growth of traffic flows to ensure that a reserve capacity for the road network being addressed can be maintained.

For other developments where large numbers of vehicle movements are unlikely, highway impacts will predominate within the traffic assessment; it will then be necessary to review junction layouts and the effect of driver delay. Assessment of queues and delays can be undertaken with a variety of software such as PICARDY and SATURN, which simulate junction interactions and downstream impacts.

Road safety is an important component of traffic assessment and an environmental assessment can be the stimulus for undertaking a review of a particular stretch of road or an existing junction. Where amendments to an existing road layout are required, it is a legal requirement for a comprehensive audit of the proposals. Some highway authorities do conduct this exercise, but increasingly this responsibility is being placed on the proponent and their consultants.

Other aspects that will require attention within the traffic assessment if they are relevant are consideration of internal site roads (developments have been refused on the basis that turning circles within a site are insufficient), provision of parking spaces and consideration of pedestrians or cycle access.

Socio-economic

There are no standard methodologies for assessing social impacts from a development; as a result, the majority of environmental statements give only scant attention to this aspect. In developing countries, however, social impacts arising from a development may be great in magnitude and this is reflected in the production of many guideline documents by the Funding Agencies such as the World Bank and the Asian Development Bank (Cernea, 1988; Asian Development Bank, 1993; Overseas Development Administration, 1993). There are indications, however, that this situation is changing as a result of awareness of the impacts that can occur as a result of a development and also through the recognition of the environmental assessment process as a tool for undertaking such studies. It is likely that this area will see significant growth in the coming years, partly because of the relative lack of guidance that currently exists and a genuine concern over the impacts that occur. Burdge (1994, 1995) and Gilpin (1995) have examined in detail the methodologies of social impact assessment and have provided valuable guidance on the solving of problems in SIA.

For economic assessment it is necessary to examine a large range of variables (depending upon the development type); one of the main problems is where to draw the boundaries for this aspect of assessment. Where large developments that are identified as having a potential socio-economic

167

impact are planned, employment of a specialist is considered essential. In reality many environmental assessments will estimate the direct socio-economic effects from a development, only broadly and then just in terms of the direct employment opportunities that may occur during identifiable phases of the development.

The aspects to include within a socio-economic assessment depend upon the size, scale and likely impacts of the development as well as on the characteristics of the area in which the development is proposed to happen. Most approaches rely upon the development and use of checklists and matrices, particularly where there are a number of different stages to the development resulting in fluctuating levels or degrees of impact (Glasson, 1995).

The aim of economic assessment is to estimate the changes in income, employment and business activity that may result from a development. As with other forms of impact assessment, it will be necessary to predict the changes that may occur on the basis of assumptions about the proposed situation in comparison that if the development does not proceed. For smaller developments where the construction phase is measured in weeks and where the workforce during the operation of the facility is either small in number or highly specialised (and is therefore brought in), the economic effects will be negligible, but will depend upon the size of the existing workforce and the rate of unemployment. For example, the operation of a landfill will perhaps generate only up to 20 jobs at a site; in an area of high unemployment, any development resulting in job vacancies is of high positive impact.

For larger developments, the direct and indirect effects will be more pronounced with substantial positive effects on the local and regional economy, for example: housing prices; the availability of associated services such as schools, hospitals and infrastructure; the employment of the local labour force; and the attraction of workers to the area. There will also be effects on the demography of the area. Negative impacts may also occur if the new development represents commercial competition to an existing facility; new jobs may be generated at the new development at the cost of redundancies at an existing and competing site. This is an aspect that is particularly difficult to quantify and it is one that is omitted from many environmental statements. A summary of potential impacts is presented in Table 14.

For some environmental assessments, the economic impacts are addressed in a purely descriptive manner through an identification of the total number of people that will be employed at the facility, what sort of employment prospects will be available (manual, blue collar,

Table 14. Potential social impacts to be assessed in EIA (adapted from Gilpin, 1995)

Social impacts (groups and aspects affected)	Micro-environmental impacts (examples of effects)
Settlement patterns	Dereliction, slums
Employment, land use	Unsafe water supply
Housing, social life, welfare	Hazards from traffic in street
Recreational and community facilities	Poorly located indusrial plants
Accessibility, safety, residential amenity	Loss of light, overshadowing of buildings
Minorities, youth unemployment, women, elderly, disabled etc.	Severence of communities and neighbourhoods by highways, railway lines etc. or large-scale development
Socio-economic profiles	
	Lack of space for play or recreation
	Visual squalor, litter, garbage vehicles
	Dereliction arising from abandoned dwellings, business premises, factories etc.
	Inadequate street maintenance and drainage
	Loss of heritage buildings and special character
	Inequitable, obtrusive, antisocial developments
	Loss of privacy
	Loss of views
	Deterioration of natural assets
	Loss of existence value

specialist and managerial) and where the jobs will be located, and an estimate of the number of houses that would need to be constructed to satisfy the housing demand.

Social impact assessment is enmeshed with the economic assessment of a development and may require inputs from anthropologists, social historians and sociologists. A large number of social impact studies have been undertaken in the USA following the incorporation of such studies within the policy-making process. The approaches to undertaking social impact assessment are broadly similar and have been summarised in a Principles document (Interorganisational Committee on Guidelines and Principles for SIA, 1994). They are:

- public consultation to involve all affected sectors of the community;
- identification of alternatives and their limitations/constraints;
- establishment of baseline conditions including examination of cultural attitudes, political and social structure, population characteristics and historical background of the area;
- evaluation of estimated effects;
- prediction of responses to impacts;
- estimation of indirect, direct and cumulative impacts;
- evaluation of alternatives or mitigation measures;
- development of mitigation measures; and
- monitoring of impacts.

The structures of socio-economic assessments are very similar to those of environmental impact assessment and the principles for undertaking the assessment are broadly the same. However there are a number of problems associated with the predictive models and assessment techniques utilised. The difficulties occur in all stages of the assessment; they are not limited to conceptual and procedural problems but also arise in the methods used.

In order to identify and assess social impacts the collation of various data and information is required, as summarised below.

- Variation and diversity within communities
  Communities are rarely homogeneous and are composed of diverse groups with different interests and stakes. These different groups need to be identified in different manners, such as:
  - occupational groups — range of employment;
  - socio-economic classification — land and capital that groups may control will differ significantly;
  - age and gender — older people may be worse affected by resettlement;
  - minority group numbers — particularly significant in undeveloped countries.
- Control over local resources
  Most local communities have some degree of control regarding environmental resources but the level of influence may vary.
- Institutions
  Decision making regarding the use of, access to and allocation of resources, and conflict resolution among competitors for resources, takes place within an institutional setting. The social impact assessment (SIA) will identify local processes contributing to the decision-making process.

- Vulnerable groups
  In order to assess impacts in this area the following information should be obtained:
    - social infrastructure — schools, medical facilities, communication networks;
    - public health conditions — health risks in an area, environmental pollution sanitation and hygiene conditions;
    - community participation — the extent to which community members feel that the development is acceptable;
    - institutional assessment — the capacity of the local authority to participate in decision making should be determined;
    - resource and area use patterns — the use of resources and the area by residents and non-residents should be analysed;
    - legal and customary use rights — these need to be determined

Models are sometimes used to predict socio-economic impacts. There are three that are most frequently used in environmental assessment:

- The Keynesian multiplier approach (Brownrigg, 1974) enables the wider economic impacts for the area, to be estimated, using an examination of the change in financial income in an area and the initial income injection and an assessment of the regional income multiplier. A number of losses are required to be taken into account which reflect the reduction of available income through such factors as taxation and savings.
- The economic base multiplier (Glasson, 1992) is a simple model that examines alterations in the local area and activities that may be dependent upon employment in a wider area. It enables a prediction of the increases in local-area employment as a function of the wider-area activities.
- Input–output models (e.g. Batey *et al.*, 1993) provide more detailed impact assessments and can reflect more accurately differences in expenditure patterns.

The models that are available that can also be used for predicting secondary effects have been reviewed elsewhere (Leistritz, 1995) and the practical problems associated with such approaches have been identified. The studies conclude that the approaches and models, as with any model, are only as good as the data used; for the model to have any reasonable accuracy large amounts of research would be necessary to address the complexity of economic interactions within an area and include consideration of such aspects as purchasing patterns, workforce requirements and capital investment in the study area, which could cover whole towns or districts. For smaller developments, and certainly for the majority of environmental assessments conducted, such a level of detail would be either impossible to collate or inappropriate for the size of development proposed.

Mitigation measures

Invariably once the impact assessment has been completed (either partially or finally), an unacceptable level of impact will be identified as likely to occur if the development were to proceed as described. The introduction of mitigation measures as part of the iterative design of the development, or less successfully as a 'bolt-on', will aim to reduce the scale and/or significance of that impact to a more acceptable level.

A recent Government publication (DETR, 1997) detailed research indicating both positive and negative aspects of the inclusion of mitigation measures in environmental statements. The research found that mitigation measures were invariably incorporated during development design rather than as an afterthought, that consideration of alternatives was addressed, that the potential for mitigation of most adverse impacts was considered, the range of mitigation measure options was limited, descriptions of mitigation were often imprecise and residual impacts were given little attention.

The incorporation of measures within the design of a development that reduces the environmental impact is an essential component of environmental assessment and one that should ideally be undertaken as an interactive part of development design rather than as an afterthought. Recent research (Mitchell, 1997) found in studies of 100 environmental statements, that treatment of mitigation was poor (or worse) in about one-third, fair in one-half and good in the remainder (18). Perhaps more worryingly, there had been no improvement in mitigation during the period studied (1990–1995). Some mitigation practices may have consequences that have little or nothing to do with their objectives (Erickson, 1994); enhancement of a borrow pit to a wetland habitat may increase waterfowl populations but may also result in a decrease in downstream water quality in groundwater wells.

Ideally, the environmental scientist tasked with advising on likely environmental impacts and responsible for undertaking or co-ordinating the environmental assessment should be invited onto the design team from the conception of the study, and will therefore be able to advise on issues and approaches that could subsequently have an environmental impact that prove difficult to mitigate. In practice, unfortunately, the majority of environmental statements are commissioned after a lengthy design period that has not involved an environmental specialist. Such an approach can often create difficulties, particularly in the event of an unacceptable impact being identified — for example, delays to the submission of the planning application, with additional expense being incurred by the design team and proponent.

Therefore mitigation should not be a reactionary response to environmental impacts that exceed statutory limits or that are unacceptable to nearby sensitive receivers, but rather it should function as a critical design tool for the development as it progresses.

Site selection represents the first opportunity for reducing environmental impacts from the proposed development. Alternative sites may not always be available (for example, the siting of an aggregate quarry, by the very nature of the geology, will preclude other sites from consideration). The 1997 amendments to the EIA Directive when implemented in member states, will require alternatives to be studied in terms of not only the location but also the process. Consideration of the 'zero option' or the 'do nothing' scenario would also be required.

Although many sites are chosen predominantly for economic, social, political or availability reasons, consideration of potential environmental impacts resulting from the development's location at the preferred site is playing a larger role in the initial planning process. Incompatible land uses are readily identified where environmental impacts from a development are likely, and significant savings can be identified in terms of both money and time if such sites are ruled out at an early stage.

Site selection exercises are particularly important for more contentious developments where some degree of impact will occur regardless of the mitigation measures that may be adopted; in such circumstances it will be necessary to confirm that a rigorous site selection study has been undertaken and the most appropriate site has been selected. The site selection exercise will be required to have a number of boundaries or constraints attached to it in order to make it an achievable task; the criteria for such developments may include geographical location and proximity to the marketplace (and to other supply/support industries), availability of services of the required capacity (water, power, sewerage, communications, transport infrastructure etc.) and the necessary physical dimensions of the required site.

One approach often used to identify the most preferable site on environmental grounds is to prepare a scoping matrix for all the sites in which the likely environmental impacts are identified so that they can be compared and the costs that will be involved to reduce these impacts can be estimated. Undertaking such an exercise in detail is invariably

173

time-consuming, costly and possibly inappropriate, in that the aim of the exercise is to provide a comparison of the impacts that may occur from all sites rather than a quantification of those impacts and the costs associated with them. The studies must be focused in their aims and objectives, to prevent them becoming lost in detail.

Once the preferred site has been selected and the environmental assessment of effects is under way, consideration of mitigation measures should commence. Reduction of impacts can be achieved through a variety of measures that target the avoidance of the impact in the first instance or enable compliance with legislative standards or preset standards; the latter may not be formal regulations but can represent accepted targets which should be met where possible. A recent report (DoE, 1995) concluded that while it was inappropriate to set definitive and absolute dust standards for dust nuisance from surface mineral workings it was acknowledged that dust guidelines will be used increasingly, particularly in environmental assessments. Reduction in the scale and magnitude of impact through the use of compensation measures that seek to replace environments or result in improvements to other areas can reflect a commitment from the developer and is an example of other general approaches to mitigation.

The mitigation measures that can be adopted to reduce the scale of impacts at a site are a function of:

- the type, scale of the development and the impacts that will occur;
- the sensitivity of the site in terms of proximity to residential properties, ecological areas of importance, water bodies etc; and
- the extent of existing impacts.

It is not possible to identify here all the mitigation measures that could be adopted as part of a development, as they will be site- and project-specific. A number of concepts have been adopted in the UK (DoE, 1990) and many measures to reduce impacts or pollution are founded on these principles. They are:

- the polluter pays;
- reduction of pollution at source;
- minimisation of risk to health (human and environmental).

These concepts are enshrined in English law through the necessity for developments (and specifically scheduled processes or pollutants as defined in the Environmental Protection Act 1990) to apply for and obtain authorisations to operate using the 'best available techniques not entailing excessive cost' (BATNEEC). Where a development is likely to produce more than one waste or effluent stream, the 'best practicable environmental option' (BPEO) must be pursued.

The incorporation of mitigation measures within a development should be based on their probable effectiveness, their appropriateness and, most importantly, their feasibility (which should be confirmed first of all). The reduction of impacts through specific measures should be designed specifically for the site, a programme for implementing the measures and achieving the reduction should be proposed (and audited) and responsibilities should be clearly defined.

Mitigation measures can take several varied forms. A recent publication (DETR, 1997) introduces a useful and practical hierarchy for mitigation:

- Avoid impacts at source.
- Reduce impacts at source.
- Abate impacts at source.
- Abate impacts at receptor.
- Repair impacts.
- Compensate in kind.
- Compensate by other means.
- Enhance.

Table 15 identifies some of the main measures that can be used to mitigate impacts. It should be noted that the examples given in the table will only be relevant in specific circumstances and in themselves can cause environmental impacts.

Although detailed, complex and wide-ranging mitigation measures may be included within an environmental statement, compliance with them does not always occur; a developer only has to follow what is described in the planning permission, and project design features such as mitigation can only be insisted upon if they are referenced in the planning permission. The reluctance of unscrupulous developers to introduce and follow 'agreed' mitigation measures is both frustrating and annoying for those involved in the environmental assessment process and can result in inclusion of very detailed conditions in the planning permission for other developments in order to prevent repetition of such faults of omission.

Many developers frequently withhold mitigation measures until the planning application has been submitted on the ground that they will be perceived as 'giving ground' to local opponents on the regulatory authorities and thus the subsequent negotiations will be easier. While this is a technique that is freqently used, recent research (DETR, 1997)

175

Table 15. Mitigation measures for environmental impacts.

Issues	Examples of mitigation measures
Water quality	Reduction/re-use/recycling of effluents
	In-house treatment systems
	Capture and diversion (and simple treatment) of run-off
	Solids and oil interceptors
	Division of water courses around developments
	Bunding of storage areas
	Maintenance of treatment systems (emptying of silt traps)
	Monitoring of effluents (before discharge)
	Protection of potential receivers (lining systems)
	Natural attenuation (reed beds)
	Direction of effluents to sewerage or specialist systems
	Maintainance of minimum river flows
	Use of buffer zones
	Habitat creation
	Maintenance of natural river systems
	Back-up safety systems (slam-shut valves etc.)
	Use of environmentally friendly techniques/methods of operation
Air quality	Reduction of vehicle speeds on construction sites
	Covering of lorries carrying earth/raw materials
	Watering on dry and windy days
	Sheeting/surrounding excavation areas
	Sweeping/cleaning of access roads
	Dust suppression systems
	High-level alarms (to storage silos etc.)
	Filters (to storage silos)
	Enclosure of conveyor belts (partial or total)
	Tree screens/windbreaks
	Enclosure of dusty activities
	Increasing height of emission stacks
	Flue-gas cleaning systems
	Increasing velocity and temperature of emissions from stacks
	Monitoring of releases at site perimeter
	Use of cleaner fuels (low-sulphur)
	Engine/combustion process efficiency and maintenance
	Burner design
	Odour-masking sprays
	Passive venting

Table 15 continued

Issues	Examples of mitigation measures
Noise	Control at source
	Anti-vibration mountings
	Regular maintenance
	Fitting of silencers
	Use of inherently quiet or alternative plant
	Non-use of reversing vehicle alarms
	Attention to location of noise source (relocation)
	Use of buffer zones
	Use of other buildings/plant to shield noisy items
	Attention to site layout
	Enclosures
	Siting works away from amenity areas
	Acoustic barriers (fences, walls, bunds, natural features)
	Receiver insulation (double glazing etc.)
	Alternative routes
	Attention to location of noise source (relocation)
	Limitation/restriction of working hours
	Avoidance of night-time working
	Provision of warnings to residents of impending activities (e.g. sirens before blast events)
Landscape	Building design (height, layout, materials, colour, shape)
	Attention to building levels (in relation to topography)
	Grouping of structures
	Attention to lighting
	Retention of natural barriers (trees, hill forms)
	Reduction in time of construction activities
	Incorporation of planting/landscaping schemes
	Screening (temporary or permanent bunds)
	Phasing of works
	Litter control
	Progressive restoration
	Relocation of species
	Replacement of landscapes (or management thereof)
	Compatibility of design with surrounding area
	Screening at point of impact (blocking of views into the site with planting measures)
	Good site management
	Reinstatement after use (temporary work sites)

Table 15 continued

Issues	Examples of mitigation measures
Ecology	Alternative sites
	Site layout (maintenance of existing habitats within site)
	Modifications to development
	Reduction of land take (optimisation of bounding location)
	Buffer zones
	Retention of specific areas of interest (quarry or natural rock faces, hedges, tree stands)
	Early restoration
	Translocation
	Restriction of access (to vehicles during construction phase)
	Maintainance of minimum river flows
	Storage of seed banks, turf etc.
	Reinstatement of habitats
	Management of non-affected habitats
	Fencing of development area to prevent encroachment
	Timing of works (to avoid breeding and nesting seasons)
	Bunding of chemical storage areas
	Sediment and oil traps
	Redirection of surface run-off from streams
	Aftercare programmes
	Compensation measures:
	provision of funding for other sites
	enhancement of on-site habitats
	agreement on management for other sites
	habitat creation
Archaeology and cultural assets	Change in alignments/location of plant
	Watching brief during construction phase
	Examination/inspection and cataloguing of remains
	Rearrangement of site layout
	Incorporation of open space in area of finds
	Preservation in situ
	Fencing
	Excavation, recording and removal of finds
	Reinstatement after use (temporary work sites)

Table 15 continued

Issues	Examples of mitigation measures
Traffic	Traffic calming
	Lighting improvement
	Junction improvement
	Carriageway and pavement widening
	Improvement of pedestrian/cycle facilities
	Improvement of sight lines
	Resurfacing
	Provision of footbridges
	Replacement of sineage
	Public transport subsidies
	Routing agreements
	Timing of delivery schedules
	Speed restrictions
	One-way routing into and from sites
	Sufficient off-street queuing and parking areas
	Financial compensation (unavoidable severance)
Socio-economic	Information dissemination (emergency plans, opportunities etc.)
	Concessions for local/new businesses
	Local or national government employment initiatives
	Skills training
	Retention of locally important cultural features
	Provision of new services/facilities
	Improvement of access
	Minimisation of land take
	Financial compensation (land take)

recognised that such submissions actually hindered the application process.

The recent review of 100 environmental statements (DETR, 1997) concluded that there was considerable scope for improvement in the treatment of mitigation measures and three areas of good practice were recommended:

- Consideration of mitigation — the full range of physical and management measures throughout the design process should be considered.
- Clarity of description — all areas of any proposed mitigation measure should be addressed, including the nature of the effect mitigated,

the location and design of the measure proposed, its effectiveness and the proposed monitoring arrangement. The document suggested inclusion of a schedule of environmental commitments (or similar) within the environmental statement; this would provide a record of the mitigation commitments made by the developer. Such an approach has been utilised by the author on a number of occasions and has been found to be very useful.

- Commitment to mitigation — clarification of the measures that will be adopted (as opposed to those that either 'should be' or 'may be' included) should be provided within the environmental statement.

Audit monitoring

Also termed environmental action plans (EAPs), environmental management plans (EMPs) or environmental response plans (ERPs), audit monitoring programmes are playing a greater and more important role with increasing frequency within the environmental assessment process, although the differences between monitoring, surveillance and survey have been identified for some time (Holdgate, 1979).

A World Bank Operational Directive refers to a project's 'mitigation plan' (EMP) which describes the mitigation measures to be taken during project implementation and operation and the actions needed to implement those measures (World Bank, 1991). Brew and Lee (1996) recognised that EMPs can play an important role in the EIA process but that the links between the two need to be strengthened. It is now fairly common for consultants to receive a scope of work that includes, in addition to the undertaking of an environmental assessment and the preparation of the environmental statement, the design and undertaking of an environmental audit monitoring exercise.

The requirement for such a programme represents the natural progression from the assessment and prediction of environmental impacts to the actual measurement of those impacts during the implementation of the development. This could be restricted to the construction phase or to the operational activities, depending upon which phase has been identified to be of most concern or where environmental impacts may be most pronounced. More probably the requirement will be for a combination of both. Audit monitoring within environmental assessment, however, is generally a weak area. This is because of:

- a lack of interest from the developers and authorities once planning permission has been obtained;
- limited legislative power to place any emphasis on the auditing phase;

- limited abilities of the monitoring activities (particularly for the 'softer' elements of environmental assessment).

Environmental assessments would appear from the outside to represent an accurate estimation of the impacts that will occur as a result of the development (not surprisingly, as this is precisely what the assessment process is attempting to achieve), but as has been discussed earlier in this chapter, the modelling and prediction of impacts — whether qualitatively or quantitatively — represents just a best guess. While the assessment processes will (or should) be based on the experience of specialists in a variety of areas, error and inaccuracies will occur regardless of the care and effort exerted. In order to quantify the impacts precisely it is necessary to measure them. This cannot be undertaken until the development commences and the impacts (with and without mitigation measures applied) become manifested. Research in Australia (Buckley, 1991) has found that predictions of impact within environmental assessment are less than 50% accurate, and over two orders of magnitude out occasionally.

The reasoning and justification for designing and undertaking audit monitoring programmes are valid and can be summarised:

- they serve as reassurance for the public that, once the development has been awarded planning permission, environmental issues will not be ignored or allowed subsequently to create potentially unacceptable impacts;
- they provide a starting point for the contract tender specification;
- they enable measurement of environmental issues (in order to manage a parameter it is necessary to define it);
- they ensure compliance with environmental legislation or pre-set environmental quality objectives;
- they ensure that measures recommended within the environmental assessment have been incorporated as part of the implementation of the project;
- they serve to gauge the accuracy of the predictions in the environmental assessment.

The scope of audit monitoring programmes will vary depending upon the potential for environmental impacts to engender concern (e.g. a large number of polluting activities on the site), the sensitivity of the surrounding environment (e.g. the proximity of the site to residential areas or the presence of a footpath that requires screening) and the perceived impacts of the development (audit monitoring can be undertaken to demonstrate a to concerned public that environmental impacts are negligible). The type, identity and number of monitoring events will be determined by these parameters while the action to be taken will depend upon the results obtained. Frameworks for conducting audits of impacts during implementation have been prepared (Barley and Fables, 1990), but often the best way for the whole of any monitoring project to be managed and controlled is as part of a certified and formalised (or even informal) environmental management system.

It is increasingly common for the operation/activities at a development to have to comply with an environmental management system (such as ISO14001 or the EMAS scheme operated on a voluntary basis by the EU) which has, as a requirement, stated goals with regard to environmental compliance and performance. It is beyond the scope of this book to review the monitoring requirements for such schemes; suffice to say that such schemes will commence with the preparation of the environmental statement in which the commitment to achieving an accredited EMS is noted. It is the author's experience that increasing numbers of environmental assessments include a draft (or outline) environmental management system. An EMS established to address environmental impacts during the construction (and operation) of a development will not just specify the policy, strategies and procedures, but also provide targets and objectives derived from the environmental assessment process and audit monitoring undertaken to date. The environmental performance of a company can then be measured and adverse environmental effects can be minimised; also, the planning authority is provided with a degree of assurance that conditions will be complied with, and where appropriate and necessary further impacts will be identified and reduced. The links between the EA process and EMS is a key area that has been identified as requiring attention for the benefit of both of these environmental management techniques (Fuller and Sadler, 1996).

The relative weakness of audit monitoring programmes within environmental assessment legislation has been identified in some countries; in Hong Kong, for example, detailed audit monitoring is a legislative requirement of EA and it is necessary for project proponents to commit to undertaking the programme at their own cost. Realistically until the UK (and EU) include an amendment to the existing controlling legislation, the value for EMPs of audit monitoring will remain limited in all but a few cases. Some regulatory authorities request that such monitoring plans are included and many consultants agree that their inclusion is both valid and useful. Because of the cost implications to the proponent, however, their adoption is limited and invariably not pursued following project completion.

Methodologies for undertaking audit monitoring vary, depending upon the sensitivity of the surrounding land uses and the specific environmental parameter concerned. Audit monitoring can be broken down into two main types of activity, each requiring a differing level of input, technical support and management. They are:

- inspection to confirm that mitigation measures that have been specified are being implemented, that terms and conditions are being met and that administrative procedures are being followed, and
- physical monitoring either to measure impact magnitude or to identify ambient environmental change.

An audit exercise will (or should) therefore highlight the accuracy of any forecasts and the performance/success of mitigation measures, and indicate the effectiveness of the monitoring or management practices (Selman, 1992). The scope and length of audits have been considered (e.g. Bisset and Tomlinson, 1988) and long periods of monitoring may be required before trends can be identified.

The audit monitoring programme must be designed with as much focus and forethought as the background monitoring programme and critically, the management tools must be in place to respond to the results obtained. These tools could consist of a collection of trigger or action levels for each parameter at which a specific action should take place. A trigger level, if exceeded, would provide an indication that a deterioration in environmental standards was occurring. The trend should be monitored closely; it may trigger a more intensive sampling programme until the levels return to normal. An action level may be set which indicates that appropriate remedial measures must be taken in order to prevent the conditions worsening further and (hopefully) to improve the environmental quality of that parameter. An example of such a situation would be the cessation of work by an earth-moving contractor as a result of high dust levels being recorded on the site perimeter. It may also be appropriate to establish maximum or target levels which would represent the absolute maximum that an environmental parameter is permitted to reach; it may correspond to a legally defined maximum (as contained in a trade effluent discharge consent). Examples of responses of exceedances of action and limit levels already used in relation to the control of construction noise have been published (Chapman, 1996).

Background monitoring programmes as part of the environmental assessment process therefore prove invaluable in establishing a level against which new impacts can be measured. Care must be taken, however, where the period of audit monitoring is so long (i.e. several years) that the background conditions change as a result of other developments taking place. Noise monitoring of a new development must take into account other unrelated activities in the near vicinity; the results of any monitoring event may not be measuring the 'target' development in isolation. An apparent exceedance of a preset limit may not be caused by the development itself but by some other factor nearby (or by a combination of factors).

The timing and programme for audit monitoring will vary in intensity throughout the phases of the development and will depend upon the results obtained as part of the environmental assessment. Baseline data will be an essential component of any audit monitoring programme. Such data are invariably collected as part of the environmental assessment but often there is need either to update or to supplement them when the development has been given approval by the planning authority. Advantage should be taken of the time between planning permission and being awarded and construction commencing on site, to collect accurate and up-to-date data sets which can be used subsequently during the development of the site. The relationship of monitoring events to the assessment process is illustrated in Fig. 12.

The type of development and the emissions and discharges it may produce will represent the main determining factor for any monitoring schemes (for example, the requirement to monitor emissions from an incinerator continuously, or the requirement for regular effluent quality samples to be taken and analysed). Other audit monitoring requirements may be defined as useful in the environmental statement, however, e.g. a yearly check on the landscaping management plan to confirm initially that all planting specified has been undertaken and in subsequent years to confirm that those landscaped areas are being managed correctly. As part of an environmental audit management programme for a solid-waste transfer

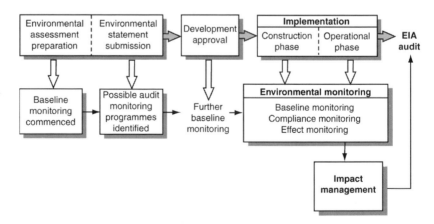

Fig. 12. Relationships and role of audit monitoring in environmental assessment

facility in Malta (Rust Environmental, 1996b), it was necessary for visits to site to be made during the construction phase to ensure that the drainage connections were being made correctly and that the suspended solids traps were being maintained correctly so that they operated at their designed efficiency. Such activities were in addition to noise- and air-monitoring programmes specified as necessary during the construction stage as part of the proposed and agreed audit programme.

In conclusion, there are a number of factors that should be adopted where possible within audit monitoring programmes in order to achieve a reasonable level of success. These are detailed below; the list cannot be all-encompassing due to the many and varied bases under which such studies may be required and the wide variety of projects (each of which will have their own site-specific nuances), but it is given as guidance for the preparation of audit monitoring programmes.

- The aim of the programme should be clearly stated and understood by all involved; this will include the brief for the monitoring team and their responsibilities, not only for undertaking the work but also with regard to reporting.
- The monitoring team should be appropriately qualified and have access to appropriate monitoring facilities and equipment.
- Where possible, the audit team should be fully independent of the construction team (difficulties will occur, however, over the definition of 'independent' and the contractual and practical problems this may cause).
- The monitoring team and programme should be able to provide competent advice on environmental issues on a site (subject to their contractual obligations); this is the best way to gain the respect of all contractors and managers on site.
- Where practical and possible, the audit monitoring team should be proactive in their response and identify potential pollution incidents before they occur.
- The monitoring team should be flexible in their approach; many larger-scale developments will be subject to changes in their programme, and the audit team should respond to such changes and be aware of the contractual, engineering and commercial constraints placed upon the contractors and operators on site whose performance they are effectively monitoring.
- Where monitoring events are occasional rather than requiring full-time presence on site, note should be made of natural fluctuations in environmental data and the results of any monitoring programme should be interpreted accordingly.

6

Case studies

AVONMOUTH CLINICAL WASTE

Introduction

This case centres around the 1993 application of Motherwell Bridge
Envirotec Ltd. (hereafter referred to as MBE) to build a clinical waste
incinerator in south-west England. To support its planning application,
MBE assessed the environmental effects of the proposed plant and
summarised its findings in an environmental statement, which was
submitted with the planning application to Avon County Council in
August 1993.

As a case study, the Avonmouth Clinical Waste Incinerator Unit is
of interest for two main reasons.

Firstly, at the time this case took place, the development of new
clinical waste incinerators was in the headlines; this was because of the
introduction of new, more stringent legislation governing the emissions
to air, water and land. This legislation was introduced in the form of the
Environmental Protection Act 1990 (EPA) and the removal of Crown
Immunity from hospitals under the National Health Service and
Community Care Act 1990. Such legislation required that current plant
used for the incineration of clinical waste would have to be replaced
with modern plant capable of meeting the new standards.

Secondly, the potential opposition of the public and of an active local
residents' committee to any additional development in the Avonmouth
area gave rise to the need to consult with the public, to allay their fears
and concerns and to demonstrate the measures that would be incorporated

within the plant design to control environmental emissions and discharges.

The project team was established by MBE in March 1993 and included:

- Matrix Environmental Ltd, a planning consultancy;
- MRM Partnership, a multidisciplinary environmental consultancy (now PB Kennedy and Donkin Ltd);
- The project proponent (MBE).

The site selection exercise had been ongoing for some 12 months prior to this; regular meetings were held by the project team with particular specialists from MBE and MRM attending as necessary.

Project proposal

MBE's proposal was for the construction of a 980 kg/h clinical waste incinerator. In addition to clinical waste incineration, the facility would incorporate a waste reception area and handling system, storage area, waste heat boiler and pollution-control equipment. The incinerator was designed to generate and be self-sufficient in its use of electricity; it would also generate a small excess that could be fed back into the National Grid.

More specifically, the proposed incinerator building would be constructed of painted profiled steel sheet to a height of 13 m at the roof apex and would cover an area of approximately 653 m². An existing single-storey office block located to the south of the site would be put to use rather than demolishing it and constructing a new building. A weighbridge for the reception of the waste would be located near the offices. The incinerator would have one 20 m stack located outside the main building. Other items adjacent to the main stack and thus outside the incinerator building would include gas-cleaning equipment, a lime storage silo and an area for ash-skip storage. An artist's impression of the proposed facility prepared as part of the environmental assessment is presented in Fig. 13.

The incinerator would dispose of clinical waste as defined by the EPA 1990 and Controlled Waste Regulations 1992. It would have a capacity of 980 kg/h and would operate 24 h a day, seven days a week. The environmental statement stated that waste would normally be brought to the site between 8 a.m. and 8 p.m. and would come mostly from what was then defined as the Avon County area, although waste from outside this catchment would also be accepted. All waste would be delivered to the site in a combination of sealed bags, boxes and plastic drums by vehicles operated by licensed waste carriers; each vehicle would have a normal capacity of 2 tonnes. Table 16 summarises the various activities that would occur at the plant.

187

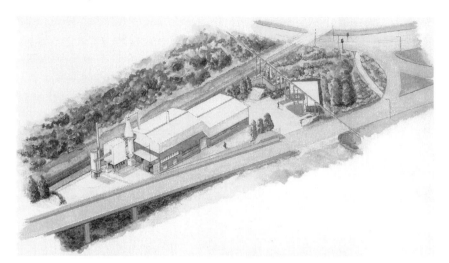

Fig. 13. Artist's impression of Avonmouth Clinical Waste Incinerator Unit (courtesy of Motherwell Bridge Envirotec Limited)

Table 16. Steps involved in the waste incineration process

Process	Activity
Reception and handling	Loading in bay
	Storage in bins
	Racking
	Transport to semi-automatic hoist (via forklift truck)
	Lifting into incinerator
Incineration	Pyrolysis at > 1000 °C
	Incineration in rotary kiln (900–950 °C)
Treatment of gaseous emissions	Water sprayed into gas stream to reduce the temperature to approximately 150 °C
	Alkaline reagent added to gas stream to react with acid gases
	Gas stream passed into fabric filter; reacted lime and particulate matter filtered out
Ash management	Removal
	Transferral to enclosed skips
	Disposal at appropriate site

Site selection and location

The main criteria for the selecting a site suitable for the proposed incinerator included the following:

- proximity to Bristol conurbation;
- location away from residential areas;
- good road access;
- existing services (water, electricity, gas etc.) with sufficient capacity.

The initial list of four potential locations was eventually narrowed down to one. Each alternative site was ruled out on the basis of one or more issues being unacceptable; the reasons were mostly environmental, arising from their proximity to residential areas (and specifically from concerns over air pollution, visual impact and ecology), but some sites were excluded on the basis of insufficient site area, poor access arrangements and site unavailability.

The location finally selected was a 0·46 ha tract of land in Avonmouth (see Fig. 14). The nearest substantial residential developments were

 Proposed
site boundary

Fig. 14. Aerial view of Avonmouth Clinical Waste Incinerator Unit (courtesy of Oldfield King Planning)

189

Avonmouth village (approximately 2 km to the south) and Lawrence Weston (2·5 km to the southeast).

One of the key features of the proposed site was its excellent transportation links: the site was only 2·9 km north of junction 18 of the M5 motorway and 10·5 km west of Bristol city centre. The improvement to the motorway and road network in the vicinity of the site had also been recently agreed (although not commenced at the time) and the site area would benefit from improvements to the local road network and through the construction of the Second Severn Crossing.

Other key features were that the site was located in an existing industrialised area on level ground with satisfactory services and the absence of conflicting land uses in the immediate vicinity.

Environmental statement

The environmental statement was prepared in accordance with the specification set out in the Town and Country Planning (Assessment of Environmental Effects) Regulations 1988 and followed the Department of the Environment guidelines. The 80-page document (not including five appendices) was divided into the following sections:

- *Non Technical Summary Statement.*
- *Introduction:* outlining the project background and providing details of relevant and related legislation.
- *Site Description:* the physical characteristics are described including site location and description, existing land use in the area, topography, flora and fauna present, and ambient air quality.
- *Proposed Development:* the incinerator and its method of operation are described in terms of layout and design; also includes waste description and handling procedures, process description, operation, emissions and monitoring procedures, waste residues and security and safety arrangements.
- *Assessment of Effects and Mitigation:* assessment of potential effects of the proposed development on the environment with sub-sections addressing air quality, solid-waste disposal, water quality and effluent disposal, visual impact, archaeological impact, ecology impact, noise impact and other impacts (including effects on humans).

In addition, a separate full-colour non-technical summary was prepared.

The contents of the environmental statement are summarised below and the main concerns of each issue are identified.

Air quality
During the construction phase, measures would be taken to ensure that dust created was kept to a minimum. Impacts were predicted to be minimal due to the limited requirement for excavations and the absence of sensitive receivers in the vicinity of the property.

During the operational phase, air quality in the vicinity of the site may be affected by combustion gases discharged via the stack. The incineration of waste would produce a range of pollutants, including hydrogen chloride, particulates, sulphur dioxide, carbon monoxide and heavy metals. To address this issue, pollution scrubbing equipment would be installed that was designed to more than satisfy the relevant emission standards.

The effect of the emissions to air on the existing background air quality was studied through the use of a computer modelling technique. The computer model determined that emissions from the incinerator during its normal operation would have a negligible impact upon existing air quality in the vicinity of the site and at the nearest residential areas and would not cause established air-quality criteria for any of the pollutants to be exceeded.

Solid-waste disposal
Solid waste produced as a by-product from the incinerator would be in the form of ash and lime residue/fly ash. This would be dampened in order to facilitate and ensure dust suppression and would be stored in enclosed skips and transported off site for disposal at a suitably licensed landfill site. Impacts on landfill capacity were determined to be of minimal significance. The waste reception and handling arrangements at the site are illustrated in Fig. 15. and the process flow diagram is presented as Fig. 16; both of these figures are taken from the environmental statement.

Water quality and effluent disposal
During facility construction it was established that there would be minimal impact upon both the foul- and surface-water drainage systems. Water usage at the site would be minimal and very little contaminated water would be produced; the construction programme was limited to a few months.

Following consultation with the relevant statutory authorities, the proposals for the disposal of both surface-water and trade effluent were agreed. Impacts upon surface water quality and the foul-sewer system from incinerator operation were considered to be minimal as a result of

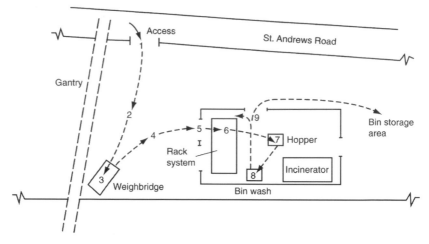

Fig. 15. Waste reception and handling arrangements at Avonmouth Clinical Waste Incinerator Unit: 1, Waste enters site in covered vehicles; 2, vehicle drives onto weighbridge; 3, vehicle weighed; 4, vehicle manoeuvres into waste reception area. 5, waste placed in wheelie bins; 6, bins conveyed to hopper via gravity-fed racking system; 7, waste deposited; 8, wheelie bin transferred to bin wash unit; 9, sterilised bins returned to storage area or for immediate re-use

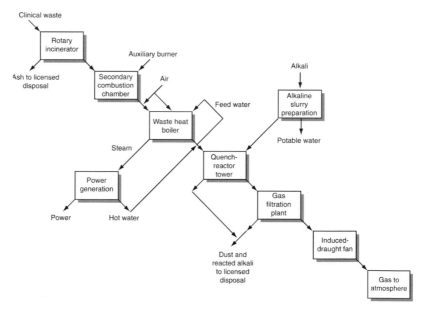

Fig. 16. Process flow diagram for Avonmouth Clinical Waste Incinerator (courtesy of Motherwell Bridge Envirotec Limited)

various mitigation measures that were incorporated into the plant design (such as use of a non-liquid gas-scrubbing system and of oil interceptors for external surface-water drain points, and provision of undercover handling and storage areas for all wastes etc.).

(a)

(b)

Fig. 17. *Photomontages of the Avonmouth Clinical Waste Incinerator Unit site (a) before construction of the Unit; (b) after its construction (courtesy of Oldfield King Planning)*

Visual impact

An assessment was made of the visual impact of the proposed development from various vantage points, including residential properties, the M5 motorway, public footpaths and other public areas.

Because of its location, i.e. in an established industrial area, the site would not have an overly negative visual impact and in the context of the surrounding environment the proposed development would be barely noticeable. As the site was in a derelict condition, the redevelopment of the site would result in a positive contribution to the character of the area. In addition, the proposed landscaping, which would include a contribution to the local 'greenway' (for pedestrians and cyclists), further impacts on the visual impact would be limited. A number of photomontages were prepared illustrating the impact of the development when constructed. These are presented in Fig. 17 and the plant following its completion is illustrated in Fig. 18.

Archaeological impact

The County Archaeologist stated that the proximity to the Bristol Channel indicated the potential for marine archaeological remains to be high. An archaeological evaluation of the site including an invasive survey was undertaken in accordance with Government Planning Policy Guidance Note 16, Archaeology and Planning. The evaluation concluded that no archaeological remains were likely to be present at the site and therefore no reason existed for withholding planning consent to develop the site.

Ecological impact

Although the proposed site was close to the Severn Estuary site of special scientific interest (SSSI), it was determined that the SSSI had no flora or fauna that would be overly sensitive to the level of emissions predicted to occur from the plant. The site itself had very little natural flora associated with it, as most of the site was laid to hardstanding.

A portion of the site (outside the fenced boundary) was covered by natural vegetation with a drainage rhyne running through it. A phase I vegetation survey reported that the area was dominated by one or two species of grass and would benefit from some positive management. Given that this area of the site would be part of the greenway running through Avonmouth, such a proposal was encouraged by the planning authority and English Nature.

Fig. 18. Avonmouth Clinical Waste Incinerator Unit after completion (courtesy of Oldfield King Planning)

Noise impact

Noise impacts during the construction of the proposed incinerator would be negligible due to the absence of any residential properties within several hundred metres of the site. The same would also hold true during the operation of the facility. Measurements of background noise levels were taken on the proposed site, and the impact assessment stated that plant operation would not significantly increase these and would certainly not result in an increase in noise levels at residential areas.

Noise produced as a result of vehicle movements delivering waste to the site for disposal was also predicted to have a minimal impact upon residential areas, particularly when the traffic management scheme for the Avonmouth area was completed. This scheme, unrelated to the proposed development, involved directing traffic away from the village of Avonmouth and would have a very beneficial effect on traffic flows and noise impacts in Avonmouth village.

The environmental statement concluded that although the development would cause some environmental impact, the scale would not be significant even under worst-case conditions.

Consultation activities

Following identification and confirmation of the proposed site in Avonmouth by MBE, consultation with Avon County Council and Bristol City Council planning departments took place almost immediately to identify any in-principle or major objections to the site proposed and proposed usage. Fears concerning the site size were raised initially but these were withdrawn following the submission of initial layout plans that illustrated sufficient turning space for vehicles, non-encumbrance of safety access and the provision of a satisfactory number of parking spaces. The local planning authority was aware that a local residents' committee was present in the area who monitored any further development within the Avonmouth area on a point of principle.

Before the planning application and environmental statement were submitted, contact was made with the residents' committee and a meeting was held at which MBE were given the opportunity to discuss their proposals and respond to questions.

This opportunity to present the proposals was taken as an ideal time to allay the concerns and apprehensions of the residents, who later commented that their opposition was mainly due to lack of knowledge and a fear of the unknown. Following the submission of the planning application for the development, no letters of opposition from nearby residents or from the Action Committee were received; indeed, one letter of support was sent by a local resident who was appreciative of the proponent's efforts to inform the residents and to reduce environmental impacts as part of the development.

The general public was not the only stakeholder to be involved in the decision-making process; the following public bodies were also consulted:

- Avon County Council Planning Department
- Bristol City Council Planning Department
- Bristol City Council Environmental Health Department
- National Rivers Authority
- South Gloucestershire Internal Drainage Board
- Her Majesty's Inspectorate of Pollution (HMIP)
- Wessex Water Authority
- Bristol City Fire Service
- Avon Waste Regulation Authority
- Bristol/Avon Regional Environmental Records Centre
- English Nature (Southwest Region)
- Avon Wildlife Trust.

Consultation with these bodies took two forms; firstly initial contact was made by telephone and letter, informing the authority of the intended development and requesting their initial comments. Secondly, a meeting was arranged with the interested bodies where they could receive more information or make particular requests for the environmental assessment to address specific issues. It was at this meeting (chaired by Avon County Council) that the scope of work was formally agreed and a programme of study was proposed.

Written correspondence with each of the authorities took place during the following assessment period; preliminary results from some of the assessment exercises were produced and issued to some of the relevant consultees for initial and informal comment.

The Bristol City Council Environmental Health Department (who would be instrumental in setting and issuing the Part B Authorisation Licence) provided invaluable advice on the format and content of the planning application and environmental statement, while the Avon Waste Regulation Authority provided useful comment and guidance on ash disposal.

Without doubt, the support from the consultees was the result of MBE's willingness to provide information and to be seen to be open with any available data and design details. This was particularly noted at the public meeting where the Head of the Waste Regulation Authority (as was) spoke, without invitation, in favour of the development

Key issues

Key issues considered by the general public as well as by the public bodies were many, but one of the most prevalent was changing legislation. Table 17 summarises the guidance notes and legislation governing the proposed incinerator.

As the table shows, both the Health and Safety Commission and the DoE advocate incineration as the preferred method of disposal for clinical waste. However, the DoE also states that landfill may be acceptable for certain types of clinical waste. To ensure that only these types of clinical waste are landfilled, however, an efficient and effective system of waste segregation is required — something which is largely unachievable, particularly within the larger hospitals.

Perhaps the most important legislation relating to the proposed incinerator is the Environmental Protection Act (EPA), which received royal assent on 1 November 1990. Whereas previous legislation was limited to incinerators with the capacity to burn 1 tonne/h or more of clinical waste, the EPA extended the system of prior control to smaller incinerators, i.e. those with the capacity to burn less than 1 tonne/h. This covers most NHS plant, as well as the proposed Avonmouth Clinical Waste Incinerator Unit.

Table 17. Guidance and legislation affecting the environmental statement assessment process

Document/Legislation	Notes
Guidance Notes	
Health and Safety Commission Guidance 1982	Recommends incineration as the preferred disposal method for most clinical waste from hospitals
DoE Waste Management Paper No. 25, 'Clinical Wastes'	Code of Practice for management, treatment and disposal of clinical waste; maintains that disposal methods other than incineration may be acceptable for certain categories
DHSS 1978 Code of Practice for the Prevention of Infection in Clinical Laboratories and Post-Mortem Rooms (the Howie Code)	All wastes falling within this code, namely all clinical waste produced from laboratories where pathological materials are examined, stored or handled, must be incinerated or autoclaved
WHO, 'Management of Waste from Hospitals', 1985	Recommends incineration in specially designed incinerators for pathological and infectious waste
DoE Waste Management Papers Nos. 14, 15 and 19	Recommendations for disposal of waste organic solvents and pharmaceuticals as special waste
Container Handling Equipment Manufacturers' Association (CHEMA) Code of Practice	For clinical-waste storage containers
London Waste Regulation Authority (LWRA) Guidelines	For segregation, handling and transport of clinical waste
National Association of Waste Disposal Contractors Guidelines	For the management of clinical waste
Legislation	
DoE Waste Management Paper No. 25 and British Standard 3316 (1973)	Detailed criteria for incineration plant handling clinical waste
Control of Industrial Pollution Regulations 1989	Regulation by HMIP of incinerators with capacity to burn 1 tonne/h or more
Environmental Protection Act 1990	Regulation of incinerators with capacity to burn less than 1 tonne/h; plant subject to system of integrated pollution control; also must comply with best available techniques not entailing excessive cost (BATNEEC) criteria

198

Table 17 continued

Document/Legislation	Notes
Process Guidance Notes 1992	Standards mandatory for plant authorisation, including chimney heights; combustion temperatures; measurement, sampling and monitoring of emissions; materials handling and storage; flue-gas treatment; disposal of solid and aqueous residues
Control of Pollution Act 1974	Provides for licensing of facilities handling or disposing of controlled wastes
Health and Safety at Work etc. Act 1974	Control of Substances Hazardous to Health Regulations 1988
Radioactive Substances Act 1993	Governs clinical waste containing radioactive substances
Toxic and Dangerous Waste Directive (78/319/EEC)	Defines what is not acceptable for landfill

Despite these new regulations — or perhaps because of them — MBE designed the new incinerator to meet the more stringent regulations governing larger plants (i.e. those burning more than 1 tonne/h), even though the capacity of the Avonmouth site was to be only 980 kg/h. This was planned with a view to future legislation, which may result in imposition of even more stringent controls on emission levels. The emission levels set out in the project proposal are presented in Table 18 (IPR 5/2) and are compared with the less stringent levels (PG 5/1) governing smaller plants.

Need

As implied in the above section on changing legislation, a strong need did indeed exist for the proposed Avonmouth Clinical Waste Incinerator Unit. The reason, quite simply, was that existing clinical waste incinerators that did not meet the new standards had been closed by October 1992, requiring waste to be transported great distances by road. The new plant was designed to meet that need and a rigorous commercial assessment was conducted by the proponent to demonstrate the viability of the Avon area as a site for an incinerator.

Table 18. Comparison of release limits for emissions to air (Guidance Notes IPR 5/2 and PG 5/1)

Emission parameter	Concentration limit (mg/m^3)	
	Plant capacity > 1 tonne/h (IPR 5/2)	Plant capacity < 1 tonne/h (PG 5/1)
Total particulate matter	30	100
VOC[a] (as carbon)	20	20
HCl	30	100
HF	2	NL[b]
SO$_2$	300	300
NO$_2$	350	NL
Dioxins[c]	1×10^{-6}	NL
Cd	0·1	—[d]
Hg	0·1	—[d]
Heavy metals (As, Cr, Cu, Pb, Mn, Ni, Sn) taken together	1·0	5·0

[a] VOC, volatile organic carbon. [b] NL, No limit specified. [c] Toxic equivalent: emission of PCDDs and PCDFs (Droxins and Furons respectively) should be reduced as much as possible by progressive techniques. The aim should be to achieve a guide TEQ value of 0·1 ng/m^3. [d] For PG 5/1, Cd and Hg are included with the other heavy metals.

Final decision and conclusions

Planning permission for the Avonmouth Clinical Waste Incineration Unit was awarded some four months after submission of the application and environmental statement in August 1993. The plant was officially opened on 7 February 1997 by the Minister of State for the Environment.

AXE VALLEY WATER RESOURCES SCHEME

Introduction

An environmental statement was prepared by MRM Partnership (now PB Kennedy and Donkin Ltd) for South West Water (SWW), addressing a water resources scheme in East Devon. The scheme consisted of a river intake and treatment works on the River Axe at Whitford, supported by a pumped storage reservoir at Higher Bruckland. The complexity of

Items ISSUED to: 7605057590

Title: Environmental assessment
ID: 7622931546
Due: 26/09/2008 23:59

Title: Sustainability indicators : measuring the
unmeasurable?
ID: 7622142959
Due: 16/05/2008 23:59

Total items: 2
19/05/2008 12:38

Thank You for using Self Service.
Please keep your receipt.

Overdue books are fined at 40p per day for
week loans, 10p per day for long loans.

the proposals and the potential scale of impacts make it an excellent subject for a case study.

The purpose of the scheme was to provide water for public supply within the East Devon area. This was to be achieved by pumping water from the River Axe to a new treatment works where it would be treated to potable standards and pumped into the supply. The scheme was also designed to provide back-up storage when river flows were too low to allow water to be abstracted from the River Axe. This storage was to be provided by means of a pumped storage reservoir and would also be available as an alternative supply in the event of river pollution upstream of the intake.

Project proposal
The proposed Axe Valley Water Resources Scheme comprised four main components:

- river intake and water treatment works on the River Axe at Whitford;
- a pumped storage reservoir (Higher Bruckland Reservoir) within a small tributary valley of the River Axe (Bruckland Stream);
- a pipeline (4 km long) connecting the river intake and treatment works within the reservoir;
- a new access route.

At times of low flow in the river, water was to be released from the reservoir to the treatment works at Whitford by gravity via the interconnecting pipeline. Releases would occur during late summer and early autumn (low flow times in the River Axe), and the reservoir would be refilled during the winter at times of high river flow by removing water from the River Axe and pumping it to Higher Bruckland.

Project team
The project team consisted of the following:

- MRM Partnership, lead consultant (engineering design and ES co-ordination) (now PB Kennedy and Donkin Ltd);
- Nicholas Pearson Associates, landscape designers;
- Dr Nigel Holmes (Alconbury Environmental Consultants) for aquatic and terrestrial ecology;
- Brooker and Garland, for water quality assessment;
- Dr. J. Alabaster, fisheries consultant;
- Dr. J. Melbourne, treatment process designer.

General report structure

The environmental statement was divided into three volumes:

- Executive summary;
- Environmental statement (with eight separate annexes addressing hydrology, water quality, fisheries, ecology, landscape, history and archaeology, recreation and consultation exercises);
- four supporting Technical Reports covering site selection, engineering, geotechnical aspects and monitoring.

The environmental statement concerned with this project was first issued in draft form to the National Rivers Authority (NRA) in April 1992. It was subdivided into four main sections and included all aspects of the proposed scheme:

- Overview of Scheme;
- River Abstraction;
- Pipeline;
- Pumped Storage Reservoir.

Each section was discussed in terms of:

- the existing situation, describing the pre-scheme situation;
- a scheme description, concerning proposed works and method of operation;
- the effects of the scheme, predicting the environmental impacts of the scheme.

The environmental statement covered:

- an assessment of the impact of the operational and construction phases;
- mitigation measures to alleviate the environmental impacts of the project;
- monitoring proposals.

The report superseded an earlier 'Environmental Assessment of River Abstraction' and Hydrological Reports (issued in November 1991 at the time of the abstraction licence applications) and was supported by a series of annexes, which consolidated studies which were carried out with respect to the scheme between 1986 and 1992, covering a number of specialist areas:

- Reservoir Site Selection — draft issued to NRA in 1991. This summarised studies of alternative reservoir sites between 1987 and 1991.

- Engineering Report — draft issued to NRA in April 1992. This consisted of a technical description of engineering considerations for the design and construction of a reservoir at Higher Bruckland and associated works.
- Geotechnical Report — draft issued to NRA in April 1992. This summarised ground investigation work made at prospective sites and conclusions concerning the geotechnical aspects of the dam and reservoir design (including groundwater and seepage aspects).
- Monitoring Report — first draft issued to NRA in May 1992.

River abstraction and treatment

Proposed scheme
The planned intake was be situated on the River Axe about 80 m upstream of Whitford Bridge. The planned water treatment works was to be situated close to the existing temporary works on the west bank of the river, immediately downstream of the bridge.

Alternative locations for the proposed intake were considered downstream of Whitford. These were subsequently rejected because of the expensive engineering works required and the probably increased visual impact. The locality selected at Whitford was considered preferable as it was relatively concealed, it was possible to utilise pre-existing weirs in the river and it was close to the proposed water treatment works. A site for the water treatment works in close proximity to the intake was preferred on landscape, cost and operational grounds.

The detailed design of the intake and water treatment works had not been completed at the time of the environmental statement, and would depend upon trial results measuring the performance of a proposed pilot treatment plant under a variety of seasonal and flow conditions. These results would determine the optimum required treatment to produce potable-quality water, and would establish whether additional treatment would be necessary before the reservoir was refilled from the river, in order to control water quality in the reservoir.

It was envisaged that the layout and architectural design of the treatment works would make maximum use of the natural advantages of the site and reflect the vernacular style, using appropriate building materials where appropriate to blend the development into the surroundings and thereby minimise visual impact.

A monitoring station was to be provided 2–3 km upstream from the intake to provide advance warning of any sudden deterioration in water quality which would force closure of the intake. For example, if accidental pollution release occurred in the river, the intake would be closed and water from the reservoir would be used until water quality in the river improved to an acceptable level for treatment.

Scheme operation

The operation of the intake would be governed by operating rules limiting when and how much water could be abstracted from the river. These rules were determined following detailed computer simulations of the proposed system to assess the effects on water levels in the reservoir and on river flows downstream of the intake point. The subsequent effects on water quality, fisheries and aquatic ecology (in the reservoir, and downstream in the river and estuary) were then assessed for a wide range of operating scenarios.

The proposed abstraction rate was 22·5 Ml/d (0·26 cumecs) or approximately 5·3% of the average daily flow in the river (4·95 cumecs). However, the river flow regularly falls to less than 1·0 cumec during the summer months. It was therefore necessary to limit the abstraction under these circumstances, in order to protect fisheries and the aquatic environment. The results of the various environmental studies indicated that there would be no detrimental effect on the river and estuary, provided that certain operating rules were observed. These included a ban on any abstraction whenever the upstream flow fell below 0·9 cumecs, and other seasonal restrictions to give added protection to salmon and sea trout during the peak migration period.

Scheme construction

A river diversion would be necessary to construct the required intake, which was estimated to take three months and timed to be carried out during the drier summer months. This was identified as the season which would have minimum impact on most aquatic life and be optimum for the engineering work, although it was realised that fish migration could be affected and that measures would be needed to avoid disturbance to the free passage of fish. Following completion of the works, the river bank would need to be restored.

Depending on the final design, construction of the water treatment works was likely to consist of earthworks (terracing and bunding), foundations, structures, installation of package plant and the laying of pipelines and services. Temporary works areas were required to accommodate spoil from the river diversion and treatment works sites. Construction access was to be limited to specific routes to ensure minimal disturbance to nearby villages.

Catchment

The River Axe at Whitford drains approximately 290 km² of rural land. The main land uses in this area are dairy and arable farming. The river water was determined to be of reasonable quality but affected by agricultural run-off, particularly at high flows. The River Axe is also recognised as an important salmonid river.

There are no major urban or industrial areas in the vicinity of the site but a large proportion of the catchment area has been assigned as an area of outstanding natural beauty (AONB), areas of great landscape value (AGLVs) and some areas include several sites of special scientific interest (SSSIs).

The catchment area comprises hilly topography with land use confined mainly to small rural settlements surrounded by woodland and agriculture. There are no large areas of open water away from the sea but the estuary is a significant feature. There are several sites of historical and archaeological interest in the valley and on surrounding hills. There are no major public supply abstractions from the Axe at present but water is abstracted from the Axe catchment, from spring sources, by South West Water and Wessex Water.

Environmental effects

Flows and water quality

Studies to assess the effects of abstraction on existing flows and water quality were carried out for the river and estuary, to establish whether:

- abstraction at Whitford would reduce the amount of river water available to dilute effluents, e.g. sewage discharges, downstream;
- reductions in flow and flow velocity might cause increased algal growth and the deposition of sediment and detritus in the river;
- reductions in the volume of fresh water might affect the relative salinity of the estuarial eco-system.

It was concluded that chemical and bacteriological quality would not be adversely affected by the abstraction. Reductions in flow and velocity were likewise predicted to be insufficient to cause a significant adverse effect.

Agricultural run-off was identified as the main cause of deterioration in river quality, particularly at times of high flow. The NRA assigned the River Axe a river quality objective (RQO) of 1B in 1991, the most recent year for which data were available at the time.

Fisheries

Assessments of the effects of proposed abstraction on fish movements concluded that minimal disturbance would occur, provided that measures were taken to protect flow conditions during critical times of the year for salmon and sea trout migration. These included additional operating rules to apply to the intake, and included limits on the amount abstracted between May and November and protection of summer spates.

The intake was designed to minimise the effects on fish and incorporated the use of 'passive' screens set into the river bed behind a permanent weir with a fish pass.

Ecology

Ecological studies included catchment-wide surveys, river corridor surveys and an estuarial survey; they covered plants, algae, invertebrates and general habitats.

The land that would be affected by the construction of the treatment works and intake site was identified as of low ecological value. The proposed construction method for the weir was designed to minimise the impact on aquatic life in the Axe, and allowed opportunity for habitat enhancement. Abstraction would cause a slight shallowing of the river, but the environmental assessment concluded that overall there would be little effect on the riverine and corridor ecology. These predictions would, however, need to be revisited after further detailed surveying and monitoring.

Landscape, history and archaeology

The improvement of the weir and restoration of the temporary diversion channel were identified as causing some local disturbance during the construction phase, but the works would ultimately result in the enhancement of the appearance of the river and banks in the area (Fig. 19). The proposed treatment works would be reasonably concealed from public view and, due to sympathetic building design and careful landscaping, visible parts of the works were not considered to be visually intrusive (Fig. 20).

Recreation and amenity

The river was identified as having limited recreational use or value; nevertheless, there was considerable tourist and conservation interest in the marsh areas adjacent to the estuary. There was some concern that the amenity could be affected by potential drops in the water level, resulting from abstraction. Water quality could subsequently be affected in the estuary and nearby beach areas by increasing concentrations of bacteria, but the effects were predicted to be insignificant for the relatively small amounts of water abstracted.

Pipeline

Design and operation
Following a detailed survey of ecological constraints and other significant features, the pipeline location was decided within a corridor 120 m wide. The proposed route would cross three lanes, a major road (the A358 from Axminster to Seaton) and the River Axe (location to be decided). The underground pipeline would be approximately 4 km long, and serve two main functions:

- to pump water to the reservoir from the intake and water treatment works;
- to allow gravity releases from the reservoir back to the water treatment works.

Environmental effects
It was intended that the area of disturbance along the pipeline route should be restored following its construction. The construction activity would cause temporary visual disturbance as it traversed the Axe Valley; however, the works would be of a short-term and temporary nature, and the route would be selected to avoid important landscape and archaeological features.

The pipelines would need to be laid across the main A358 (Axminster to Seaton) road as well as the road from Musbury to Whitford. This operation would involve either trenching or possibly thrust-boring (to minimise traffic disruption).

The pipelines would cross many fields, but its final route would be sited to avoid specific areas of hedgerow and individual trees of high ecological interest, although some long-term impact would be unavoidable.

Pumped storage reservoir

Proposed scheme
Several different options were considered for reservoir storage as part of the environmental assessment. These included a bankside reservoir

207

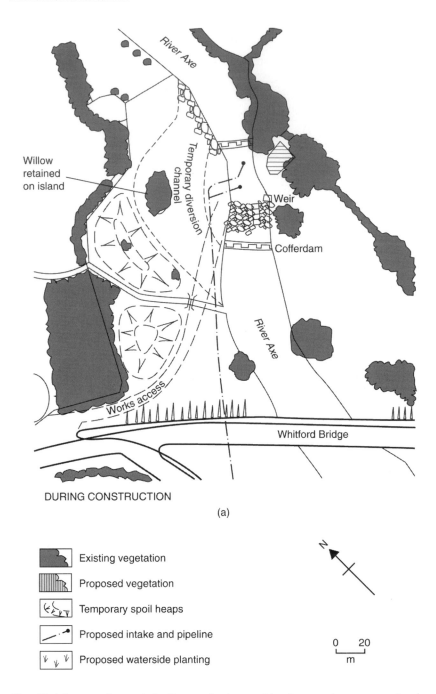

Willow
retained
on island

Temporary diversion channel

River Axe

Weir

Cofferdam

River Axe

Works access

Whitford Bridge

DURING CONSTRUCTION

(a)

Existing vegetation

Proposed vegetation

Temporary spoil heaps

Proposed intake and pipeline

Proposed waterside planting

N

0   20
      m

Fig. 19 (above and opposite). Proposed scheme of landscape reinstatement for the river intake of the Axe Valley Water Resource Scheme (courtesy of Nicholas Pearson Associates Ltd)

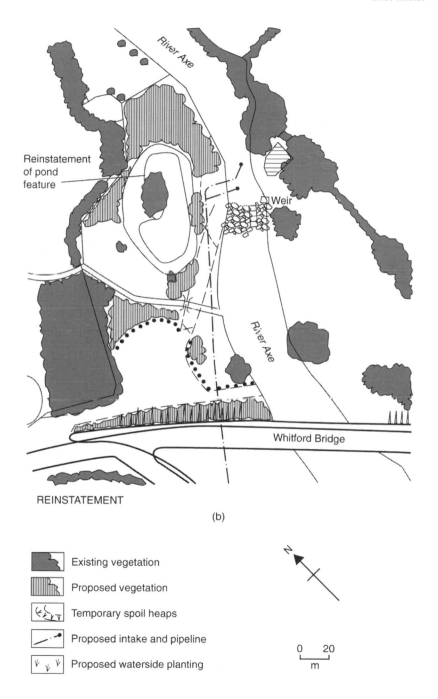

Reinstatement
of pond
feature

Weir

River Axe

River Axe

Whitford Bridge

REINSTATEMENT

(b)

Existing vegetation

Proposed vegetation

Temporary spoil heaps

Proposed intake and pipeline

Proposed waterside planting

N

0   20
    m

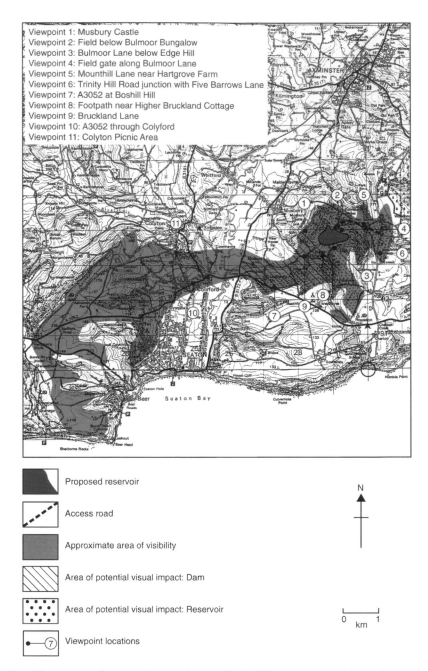

Viewpoint 1: Musbury Castle
Viewpoint 2: Field below Bulmoor Bungalow
Viewpoint 3: Bulmoor Lane below Edge Hill
Viewpoint 4: Field gate along Bulmoor Lane
Viewpoint 5: Mounthill Lane near Hartgrove Farm
Viewpoint 6: Trinity Hill Road junction with Five Barrows Lane
Viewpoint 7: A3052 at Boshill Hill
Viewpoint 8: Footpath near Higher Bruckland Cottage
Viewpoint 9: Bruckland Lane
Viewpoint 10: A3052 through Colyford
Viewpoint 11: Colyton Picnic Area

Proposed reservoir

Access road

Approximate area of visibility

Area of potential visual impact: Dam

Area of potential visual impact: Reservoir

Viewpoint locations

Fig. 20. Reservoir location plan for the Axe Valley Water Resource Scheme (courtesy of Nicholas Pearson Associates Ltd; based upon the 1985 Ordnance Survey 1:50 000 Landranger map with permission of the Controller of Her Majesty's Stationery Office, © Crown Copyright MC 88595M0001)

situated adjacent to the River Axe at Whitford or a barrage situated at the top end of the estuary, but these ideas were rejected at an early stage because of the relatively large area required, and the significant intrusion into the landscape. The barrage also had considerable implications for both water quality and recreation. In addition, large volumes of material would have to be excavated from the valley floor and disposed of to create the required storage volume.

The preferred option was for the construction of a reservoir in a small tributary valley of the River Axe and this was subsequently considered in more detail. Water pumped from the river would be stored behind an embankment dam constructed from in-situ material removed from the valley floor and sides to help create the lake.

Seven sites were identified as possible dam locations, but following initial studies (1987) the number of possible sites was reduced to five. During subsequent work (1988–1989) more detailed geotechnical studies were made, involving borehole drilling and trial pit excavation. This established the suitability of ground conditions. This latter stage also involved a more detailed assessment of storage volume requirements, resulting in the inclusion of one additional site (which had a greater storage potential). As a result, three possible locations were shortlisted for a further stage of study.

The most recent studies (1990–1991) involved more detailed environmental and geotechnical assessment. Following this stage, the site at Higher Bruckland (Fig. 21) was chosen as the preferred dam site. This site offered a number of major benefits, including:

- it had more favourable topography and soil conditions;
- application of materials lining to the sides and base of the reservoir to reduce seepage losses would not be unduly difficult;
- there was greater potential for integrating the dam and reservoir into the valley setting.

A number of adverse features were also recognised:

- the requirement for new road access;
- intrusion into a quiet valley;
- proximity to an important SSSI;
- loss of agricultural land.

New access routes would be required; they were to be confined to public highways where possible and it was stated that construction traffic should be confined to the permanent access route to the site in order to minimise impacts. The route chosen had good connections to the A35

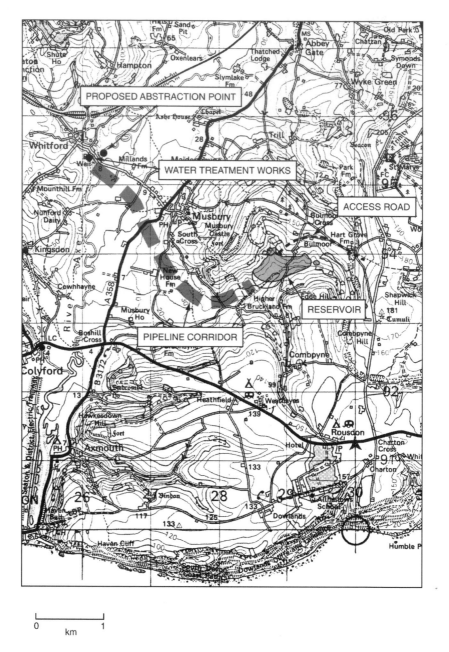

Fig. 21. Site visibility of the reservoir and access road for the Axe Valley Water Resource Scheme (based upon the 1985 Ordnance Survey 1:50 000 Landranger map with permission of the Controller of Her Majesty's Stationery Office, © Crown Copyright MC 88595M0001)

and A3052 coastal route, it had optimum alignment with respect to directness and gradient, and it would avoid damage to adjacent mature trees and meadows.

Scheme description

The proposed reservoir scheme had the following characteristics which were assessed as part of the environmental statement:

- main plan dimensions 750 m × 350 m;
- Surface area 22·5 ha (when full);
- Storage volume 2850 Ml (when full);
- Water impoundment behind a main embankment dam and a smaller 'backstop' dam;
- 1 km of existing stream and 30 ha of farmland lost due to construction;
- the stream would be diverted around the reservoir, as the quality was not suitable for reservoir impoundment, and it was therefore possible to conserve the water course downstream of the reservoir;
- water would be pumped from Whitford under suitable river flow conditions, and drawn off through the main dam for return to Whitford under low flow conditions in the Axe;
- an access road to the reservoir for operational purposes and for leisure amenity would be constructed and a visitors' car park was also planned;
- the access road would follow some existing routes — road improvements would be necessary and care would have to be taken to avoid damage to the environment along the newly built sections;
- the main dam construction would exploit the topography and was shaped to harmonise with the landscape — it was to be constructed from naturally occurring clays and marls from the area within the reservoir basin;
- the backstop dam would limit the submergence area of the reservoir at the eastern end, thus protecting a small SSSI at the upstream end;
- the borrow area would be submerged following reservoir filling, and the cut slopes above the water level would be landscaped and restored;
- basin treatment would be applied, involving the lining of the base and sides with clay and marls to minimise water loss by underground seepage;
- an overflow spillway would be built to discharge any excess water due to rainfall on the reservoir surface;
- low-level drawoff was provided to allow total drawdown of the reservoir if dam repairs were required;

- public amenities were to be provided, which could possibly become a tourist attraction, and would be designed in a manner which will minimise visual impact;
- the lake would be developed as a coarse fishery and a recreational amenity;
- it was hoped to encourage and promote positive environmental management of land immediately adjacent to the lake to provide an environmental haven.

Reservoir construction

The construction period was estimated to be approximately two years:

- Before construction started, vegetation and mammals would be relocated and archaeological excavation would be instigated.
- Dam fill materials would be obtained from the site, but other construction materials would require importation by road. The level of traffic generated by the site works was predicted to vary over the construction period, being heaviest in the first nine months.
- Agreement had been obtained from The Highway Authority that all traffic should be routed onto the site from the A35. Other local routes would require some upgrading, as the contractor was forbidden from using any other public access to the site.
- There was likely to be an increased risk of pollution to surface and groundwater during the construction period. During construction the stream would be diverted and run-off from the site would be contained within temporary lagoons to settle out the suspended solids prior to discharge to the stream.
- Control and supervision of construction had been carried out in such a manner as to ensure the minimum effect on the SSSI.
- To ensure that there was no potential for ecological degradation, a careful assessment of locations for soil disposal from the reservoir had to be made.
- Visual impact during the general construction period was inevitable, producing a temporary disturbance to the valley.
- As public interest would be generated by the construction, visitor management policies had to be adopted to increase public awareness and to protect the site.

Reservoir construction has many potential significant additional impacts. Although the construction stage is of a temporary nature, resulting impacts vary in severity and duration. Some impacts are immediately detectable; others may become apparent only after a considerable time has elapsed.

Reservoir operation

The reservoir was required as a reserve supply when low flow in the River Axe prevented normal abstraction. It would also provide a back-up supply in the event of pollution or high sediment loads in the Axe under flood conditions. The amount of drawdown would vary annually.

The Axe at Whitford was suitable for potable supply following treatment. The river was eutrophic, so algal growth during storage would need to be prevented either by pretreatment at Whitford or circulation of reservoir water. The proposed maximum depth of the reservoir exceeded 20 m, so seasonal stratification of water quality was likely. It was intended to introduce aeration equipment to circulate the water to prevent poor-quality water from collecting at the reservoir bottom and to inhibit algal growth near the surface.

The reservoir water would be treated at Whitford before supply, and modelling would be necessary to establish the required treatment process. However, it was predicted that some self-purification will occur because of the long storage times, resulting in improvement in water quality prior to treatment.

Environmental effects

Climate and drawdown

The introduction of a new water body to the valley may have a slight localised effect on the valley microclimate, and may be beneficial; no adverse effects were expected. In dry years, drawdown of the reservoir would supply the demands that could not be supplied by the river. The drawdown would vary annually, but it has been estimated that a drawdown of, on average, 0·5 m would be needed every two years. However, this would be increased by additional use for pollution control. A reasonable estimate of annual average fluctuations is approximately 1–2 m.

Geology and slope stability

The reservoir is underlain by clays, marls and mudstones of the Mercia Mudstone Group, which are in turn overlain by the Upper Greensand on upper slopes. No faults were encountered during the ground investigations, but a number of minor faults were identified in the area. These faults posed no great problems to dam construction on the site. Evidence was noted in several areas of remnant ancient mudflows, implying that shallow landslips had occurred in the past.

It was predicted that the reservoir would change the natural groundwater pattern locally, but that it would not induce deep-seated slope-stability problems. The east side of the reservoir contains current and progressive instability areas, which are poorly drained and are fed by spring water from the Greensand above, forming part of the new SSSI. The backstop dam was designed to protect the unstable area from fluctuating water levels in the main water body. Ponding of the Bruckland Stream on the upstream side would produce a constant water level; no drainage was proposed for this area as it would alter the character of the groundwater-dependent SSSI habitat. New vegetation would be planted in the area to improve the appearance and provide erosion protection of the shoreline.

Groundwater
The regional water table is located within the mudstone and perched in the Greensand in slopes above the reservoir, producing the wetland features in the SSSI. The reservoir would be lined with clay to minimise seepage losses and it was not anticipated that significant changes in the volume of groundwater reaching the stream would occur. Local groundwater patterns would be affected and it was thought they might produce localised new spring areas downstream of the dam, but studies showed these to be of minimal significance and impact.

Surface water
Stream flows were predicted to be little affected by the reservoir, but it was suggested that small decreases in surface run-off and possible increases in baseflow downstream of the dam due to steady reservoir seepage under low flow conditions might occur. Water quality would be unlikely to be significantly affected under normal operation. However, possible adverse effects were:

- Seepage of poor-quality water from the reservoir base. This could and should be alleviated by the presence of the clay liner and long drainage path through the underlying soils.
- Release of small volumes of poor-quality water from the depths of the reservoir during routine testing once a year. It was intended that released water should be temporarily stored downstream from the dam, and subsequently released slowly into the stream or passed down to Whitford for release into the Axe.

- Release of water from the access road and car parks. This would be diverted around the reservoir and into the stream, but such drainage would be subject to mitigation measures to protect the water course (silt and oil traps etc.).

Ecology
NRA and South West Water surveys indicated that aquatic life in the stream was typical of others in the area and not of special interest. Nevertheless, the ecology was considered of sufficient importance to justify protection from potential damage due to the scheme. A change in the flow regime was predicted not to cause a significant long-term impact, but short-term construction impacts were identified as potentially more damaging. The diversion culvert and spillway would affect fish and animals and result in a loss of stream habitat. This would be unavoidable but would be compensated by the creation of a new coarse fishery in the reservoir and a new pond habitat behind the backstop dam.

Terrestrial ecology
Surveys revealed that sites of high ecological interest did exist in the vicinity of the site, but that they were found predominantly outside the area to be submerged by the reservoir, which was sited partly to avoid them. There were no habitats of any special interest along the proposed access route, which was specifically aligned to avoid areas of interest. Specific measures would be included in the road design to ensure that natural groundwater flow to the wetland was not affected.

Landscape, history and archaeology
The rural nature of the landscape is valued by local residents and the users of public rights of way through the valley, so it was stipulated that the landscape importance of the site (within an AONB) had to be accounted for in the design. There would be a loss of some features, but in compensation there would be the addition of a new element to the landscape with new wildlife habitats and informal countryside recreation. A detailed survey was recommended within the inundation zone to ensure that all archaeological features were identified.

Land use and water use
The land in the catchment area was predominantly agricultural, comprising grassland, arable cropping and grazing by sheep and cattle. Some farming practices were identified as posing a potential pollution risk to the reservoir, but this could be prevented by installing perimeter

drains which would intercept surface run-off, while streams would be diverted around the reservoir.

Licensed abstractions exist for wells and springs in the area, but it was thought possible that two nearby sources could be affected by the access road, and would therefore require some form of protection. Mitigation methods considered included:

- provision of a mains supply;
- redrilling wells north of the road alignment;
- minor re-routing of the road to the south.

Recreation and amenity

The reservoir would be located within the West Country tourist market. There would be positive effects on recreation through the provision of a new lake and coarse fisheries. Other leisure facilities might be considered but these would need to be appropriate to the peaceful setting of the valley. The valley was currently used for low-key recreational activities such as rambling and horse-riding. Local inhabitants were most interested in maintaining the seclusion, tranquillity and nature conservation aspects of the landscape.

Consultation

A lengthy consultation process was carried out as part of the outline design of the scheme and preparation of the environmental statement. The organisations and groups consulted included:

- East Devon District Council (planning authority)
- Devon County Council (highway authority)
- National Rivers Authority
- Countryside Commission
- English Nature
- local wildlife trusts
- parish councils
- landowners and the public
- action groups
- angling bodies

The above list is by no means exhaustive.

The planning authority, while expressing considerable interest in the potential recreational benefits, reserved their position until such time as a planning application was submitted and formal consultations sought.

Agreement was obtained in principle with the highways authority for the route of the proposed access road to the reservoir, including access for construction traffic.

Conservation bodies were in general opposed to the siting of a new development in an AONB and in such close proximity to SSSIs. It was considered that a case had to be made that there was an overriding need for the reservoir and reasons given for this choice of strategy. Wildlife trusts and angling clubs had concerns over the possible impact of the abstraction on the lower river and estuary.

The NRA were concerned primarily with the scheme justification within the context of the overall strategy for water supply to east Devon, although their concerns also encompassed the impact on both the main river and tributary catchment environments.

Landowners included those directly affected by loss of land or land severance, and those indirectly affected. Landowners were particularly concerned at the intrusion of a new reservoir, however sympathetically designed, and the potential number of visitors to the relative seclusion of the Bruckland valley. There was less concern over the treatment works proposals.

The public were divided in opinion between those who understood the reasoning behind a new source of water to make good the summer water shortages and those who were dismayed over the potential environmental impact.

Status of application

Full documentation was prepared in support of a planning application and concurrent abstraction and impoundment licence applications from the NRA (now the Environment Agency). Approval in principle was sought from the NRA before submitting any planning application.

In the event, the NRA had a number of fundamental concerns over regional water resources strategy, including demand forecasts, the amount allowed for leakage reduction and other factors. Following a strategy review, it was decided that sufficient water could be obtained by transfer from the River Exe and that subsequent demand increases could be covered by promoting an extension to the Wimbleball Reservoir storage scheme. It was therefore decided not to proceed with the Axe Valley Water Resource Scheme in the foreseeable future.

Appendix 1

The text of Council Directive of 27 June 1985
on the assessment of the effects of certain public and private projects
on the environment 85/337/EEC

THE COUNCIL OF THE EUROPEAN COMMUNITIES,

Having regard to the Treaty establishing the European Economic Community, and in particular Articles 100 and 235 thereof,

Having regard to the proposal from the Commission,

Having regard to the opinion of the European Parliament,

Having regard to the opinion of the Economic and Social Committee,

Whereas the 1973 and 1977 action programmes of the European Communities on the environment, as well as the 1983 action programme, the main outlines of which have been approved by the Council of the European Communities and the representatives of the Governments of the Member States, stress that the best environmental policy consists in preventing the creation of pollution or nuisances at source, rather than subsequently trying to counteract their effects; whereas they affirm the need to take effects on the environment into account at the earliest possible stage in all the technical planning and decision-making processes; whereas to that end, they provide for the implementation of procedures to evaluate such effects;

Whereas the disparities between the laws in force in the various Member States with regard to the assessment of the environmental effects of public and private projects may create unfavourable competitive conditions and thereby directly affect the functioning of the common market; whereas, therefore, it is necessary to approximate national laws in this field pursuant to Article 100 of the Treaty;

Whereas, in addition, it is necessary to achieve one of the Community's objectives in the sphere of the protection for the environment and the quality of life;

Whereas, since the Treaty has not provided the powers required for this end, recourse should be had to Article 235 of the Treaty;

Whereas general principles for the assessment of environmental effects should be introduced with a view to supplementing and co-ordinating development consent procedures governing public and private projects likely to have a major effect on the environment;

Whereas development consent for public and private projects which are likely to have significant effects on the environment should be granted only after prior assessment of the likely significant environmental effects of these projects has been carried out; whereas this assessment must be conducted on the basis of the appropriate information supplied by the developer, which may be supplemented by the authorities and by the people who may be concerned by the project in question;

Whereas the principles of the assessment of environmental effects should be harmonised, in particular with reference to the projects which should be subject to assessment, the main obligations of the developers and the content of the assessment;

Whereas projects belonging to certain types have significant effects on the environment and these projects must as a rule be subject to systematic assessment;

Whereas projects of other types may not have significant effects on the environment in every case and whereas these projects should be assessed where the Member States consider that their characteristics so require;

Whereas, for projects which are subject to assessment, a certain minimal amount of information must be supplied, concerning the project and its effects;

Whereas the effects of a project on the environment must be assessed in order to take account of concerns to protect human health, to contribute by means of a better environment to the quality of life, to ensure maintenance of the diversity of species and to maintain the reproductive capacity of the ecosystem as a basic resource for life;

Whereas, however, this Directive should not be applied to projects the details of which are adopted by a specific act of national legislation, since the objectives of this Directive, including that of supplying information, are achieved through the legislative process;

Whereas, furthermore, it may be appropriate in exceptional cases to exempt a specific project from the assessment procedures laid down by this Directive, subject to appropriate information being supplied to the Commission,

HAS ADOPTED THIS DIRECTIVE

Article 1

1. This Directive shall apply to the assessment of the environmental effects of those public and private projects which are likely to have significant effects on the environment.

2. For the purposes of this Directive:
    "project" means:

    (i) the execution of construction works or of other installations or schemes,
    (ii) other interventions in the natural surroundings and landscape including those involving the extraction of mineral resources;

    "developer" means:
    the applicant for authorisation for a private project or the public authority which initiates a project;
    "development consent" means:
    the decision of the competent authority or authorities which entitles the developer to proceed with the project.

3. The competent authority or authorities shall be that or those which the Member States designate as responsible for performing the duties arising from this Directive.

4. Projects serving national defence purposes are not covered by this Directive.

5. This Directive shall not apply to projects the details of which are adopted by a specific act of national legislation, since the objectives of this Directive, including that of supplying information, are achieved through the legislative process.

Article 2

1. Member states shall adopt all measures necessary to ensure that, before consent is given, projects likely to have significant effects on the environment by virtue *inter alia*, of their nature, size or location are made subject to an assessment with regard to their effects.

These projects are defined in Article 4.

2. The environmental impact assessment may be integrated into the existing procedures for consent to projects in the Member States, or, failing this, into other procedures or into procedures to be established to comply with the aims of this Directive.

3. Member States may, in exceptional cases, exempt a specific project in whole or in part from the provisions laid down in this Directive.
    In this event, the Member States shall:

    (a) consider whether another form of assessment would be appropriate and whether the information thus collected should be made available to the public;

    (b) make available to the public concerned the information relating to the exemption and the reasons for granting it;

    (c) inform the Commission, prior to granting consent, of the reasons justifying the exemption granted, and provide it with the information made available, where appropriate, to their own nationals.

The Commission shall immediately forward the documents received to the other Member States.

The Commission shall report annually to the Council on the application of this paragraph.

Article 3

The environmental impact assessment will identify, describe and assess in an appropriate manner, in the light of each individual case and in accordance with the Articles 4 to 11, the direct and indirect effects of a project on the following factors:

—   human beings, fauna and flora,

—   soil, water, air, climate and the landscape,

—   the inter-action between the factors mentioned in the first and second indents,

—   material assets and the cultural heritage.

Article 4

1. Subject to Article 2(3), projects of the classes listed in Annex I shall be made subject to an assessment in accordance with Articles 5 to 10.

2. Projects of the classes listed in Annex II shall be made subject to an assessment, in accordance with Articles 5 to 10, where Member States consider that their characteristics so require. To this end, Member States may *inter alia* specify certain types of projects as being subject to an assessment

or may establish the criteria and/or thresholds necessary to determine which of the projects of the classes listed in Annex II are to be subject to an assessment in accordance with Articles 5 to 10.

Article 5

1. In the case of projects which, pursuant to Article 4, must be subjected to an environmental impact assessment in accordance with Articles 5 to 10, Member States shall adopt the necessary measures to ensure that the developer supplies in an appropriate form the information specified in Annex III inasmuch as:

    (a) The Member States consider that the information is relevant to a given stage of the consent procedure and to the specific characteristics of a particular project or type of project and of the environmental features likely to be affected;

    (b) the Member States consider that a developer may reasonably be required to compile this information having regard *inter alia* to current knowledge and methods of assessment.

2. The information to be provided by the developer in accordance with paragraph 1 shall include at least:

    — a description of the project comprising information on the site, design and size of the project,

    — a description of the measures envisaged in order to avoid, reduce and, if possible, remedy significant adverse effects,

    — the data required to identify and assess the main effects which the project is likely to have on the environment,

    — a non-technical summary of the information mentioned in indents 1 to 3.

3. Where they consider it necessary, Member States shall ensure that any authorities with relevant information in their possession make this information available to the developer.

Article 6

1. Member States shall take the measures necessary to ensure that the authorities likely to be concerned by the project by reason of their specific environmental responsibilities are given an opportunity to express their

opinion on the request for development consent. Member States shall designate the authorities to be consulted for this purpose in general terms or in each case when the request for consent is made. The information gathered pursuant to Article 5 shall be forwarded to these authorities. Detailed arrangements for consultation shall be laid down by the Member States.

2.   Member States shall ensure that:

(i)   any request for development consent and any information gathered pursuant to Article 5 are made available to the public,

(ii) the public concerned is given the opportunity to express an opinion before the project is initiated.

3.   The detailed arrangements for such information and consultation shall be determined by the Member States, which may in particular, depending on the particular characteristics of the projects or sizes concerned:

(i)   determine the public concerned,

(ii) specify the places where the information can be consulted,

(iii) specify the way in which the public may be informed, for example, by bill posting within a certain radius, publication in local newspapers, organisation of exhibitions with plans, drawings, tables, graphs, models,

(iv) determine the manner in which the public is to be consulted, for example by written submissions, by public enquiry,

(v)   fix appropriate time limits for the various stages of the procedure in order to ensure that a decision is taken within a reasonable period.

Article 7

Where a Member State is aware that a project is likely to have significant effects on the environment in another Member State or where a Member State likely to be significantly affected so requests, the Member State in whose territory the project is intended to be carried out shall forward the information gathered pursuant to Article 5 to the other Member State at the same time as it makes it available to its own nationals. Such information shall serve as a basis for any consultations necessary in the framework of the bilateral relations between two Member States on a reciprocal and equivalent basis.

Article 8

Information gathered pursuant to Articles 5, 6 and 7 must be taken into consideration in the development consent procedure.

Article 9

When a decision has been taken, the competent authority or authorities shall inform the public concerned of:

(i) the content of the decision and any conditions attached thereto,

(ii) the reasons and consideration on which the decision is based where the Member States' legislation so provides.

The detailed arrangements for such information shall be determined by the Member States.

If another Member State has been informed pursuant to Article 7, it will also be informed of the decision in question.

Article 10

The provisions of this Directive shall not affect the obligation on the competent authorities to respect the limitations imposed by national regulations and administrative provisions and accepted legal practices with regard to industrial and commercial secrecy and the safeguarding of the public interest.

Where Article 7 applies, the transmission of information to another Member State and the reception of information by another Member State shall be subject to the limitations in force in the Member State in which the project is proposed.

Article 11

1. The Member States and the Commission shall exchange information on the experience gained in applying this Directive.

2. In particular, Member States shall inform the Commission of any criteria and/or thresholds adopted for the selection of the projects in question, in accordance with Article 4(2), or of the types of projects concerned which, pursuant to Article 4(2), are subject to assessment in accordance with Articles 5 to 10.

3. Five years after notification of this Directive, the Commission shall send the European Parliament and the Council a report on its application and effectiveness. The report shall be based on the aforementioned exchange of information.

4. On the basis of this exchange of information, the Commission shall submit to the Council additional proposals, should this be necessary, with a view to this Directive's being applied in a sufficiently co-ordinated manner.

Article 12

1. Member States shall take the measures necessary to comply with this Directive within three years of its notification.

2. Member States shall communicate to the Commission the texts of the provisions of national law which they adopt in the field covered by this Directive.

Article 13

The provisions of this Directive shall not affect the right of Member States to lay down stricter rules regarding scope and procedure when assessing environmental effects.

Article 14

This Directive is addressed to the Member States.

Done at Luxembourg, 27 June 1985.

For the Council

The President

A. BIONDI

ANNEX I
PROJECTS SUBJECT TO ARTICLE 4(1)

1. Crude-oil refineries (excluding undertakings manufacturing only lubricants from crude oil) and installations for the gasification and liquefaction of 500 tonnes or more of coal or bituminous shale per day.

2. Thermal power stations and other combustion installations with a heat output of 300 megawatts or more and nuclear power stations and other nuclear reactors (except research installations for the production and conversion of fissionable and fertile materials, whose maximum power does not exceed 1 kilowatt continuous normal load).

3. Installation solely designed for the permanent storage or final disposal of radioactive waste.

4. Integrated works for the initial melting of cast-iron and steel.

5. Installations for the extraction of asbestos and for the processing and transformation of asbestos and products containing asbestos: for asbestos-cement products, with an annual production of more than 20,000 tonnes of finished products, for friction material, with an annual production of more than 50 tonnes of finished products, and for other uses of asbestos, utilisation of more than 200 tonnes per year.

6. Integrated chemical installations.

7. Construction of motorways, express roads and lines for long distance railway traffic and of airports with a basic runway length of 2,100 metres or more.

8. Trading ports and also inland waterways and ports for inland-waterway traffic which permit the passage of vessels of over 1350 tonnes.

9. Waste-disposal installations for the incineration, chemical treatment or land fill of toxic and dangerous waste.

ANNEX II
PROJECTS SUBJECT TO ARTICLE 4(2)

1. *Agriculture*

    (a) Projects for the restructuring of rural land holdings;

    (b) Projects for the use of uncultivated land or semi-natural areas for intensive agricultural purposes;

(c) Water-management projects for agriculture;

(d) Initial afforestation where this may lead to adverse ecological changes and land reclamation for the purposes of conversion to another type of land use;

(e) Poultry-rearing installations;

(f) Pig-rearing installations;

(g) Salmon breeding;

(h) Reclamation of land from the sea.

2. *Extractive industry*

    (a) Extraction of peat;

    (b) Deep drillings with the exception of drillings for investigating the stability of the soil and in particular;

    — geothermal drilling

    — drilling for the storage of nuclear waste material

    — drilling for water supplies

    (c) Extraction of minerals other than metalliferous and energy-producing minerals, such as marble, sand, gravel, shale, salt, phosphates and potash;

    (d) Extraction of coal and lignite by underground mining;

    (e) Extraction of coal and lignite by open-cast mining;

    (f) Extraction of petroleum;

    (g) Extraction of natural gas;

    (h) Extraction of ores;

    (i) Extraction of bituminous shale;

    (j) Extraction of minerals other than metalliferous and energy-producing minerals by open-cast mining;

(k) Surface industrial installations for the extraction of coal, petroleum, natural gas and ores, as well as bituminous shale;

(l) Coke ovens (dry coal distillation);

(m) Installations for the manufacture of cement.

3. *Energy industry*

(a) Industrial installations for the production of electricity, steam and hot water (unless included in Annex I);

(b) Industrial installations for carrying gas, steam and hot water; transmission of electrical energy by overhead cables;

(c) Surface storage of natural gas;

(d) Underground storage of combustible gases;

(e) Surface storage of fossil fuels;

(f) Industrial briquetting of coal and lignite;

(g) Installations for the production or enrichment of nuclear fuels;

(h) Installations for the reprocessing of irradiated nuclear fuels;

(i) Installations for the collection and processing of radioactive waste (unless included in Annex I);

(j) Installations for hydroelectric energy production.

4. *Processing of metals*

(a) Iron and steelworks, including foundries, forges, drawing plants and rolling mills (unless included in Annex I);

(b) Installations for the production, including smelting, refining, drawing and rolling, of non-ferrous metals, excluding precious metals;

(c) Pressing, drawing and stamping of large castings;

(d) Surface treatment and coating of metals;

(e) Boilermaking, manufacture of reservoirs, tanks and other sheet-metal containers;

(f) Manufacture and assembly of motor vehicles and manufacture of motor-vehicle engines;

(g) Shipyards;

(h) Installations for the construction and repair of aircraft;

(i) Manufacture of railway equipment;

(j) Swaging by explosives;

(k) Installations for the roasting and sintering of metallic ores.

5. *Manufacture of glass*

6. *Chemical industry*

   (a) Treatment of intermediate products and production of chemicals (unless included in Annex I);

   (b) Production of pesticides and pharmaceutical products, paint and varnishes, elastomers and peroxides;

   (c) Storage facilities for petroleum, petrochemical and chemical products.

7. *Food industry*

   (a) Manufacture of vegetable and animal oils and fats;

   (b) Packing and canning of animal and vegetable products;

   (c) Manufacture of dairy products;

   (d) Brewing and malting;

   (e) Confectionery and syrup manufacture;

   (f) Installations for the slaughter of animals;

   (g) Industrial starch manufacturing installations;

(h) Fish-meal and fish-oil factories;

(i) Sugar factories.

8. *Textile, leather, wood and paper industries*

(a) Wool scouring, degreasing and bleaching factories;

(b) Manufacture of fibre board, particle board and plywood;

(c) Manufacture of pulp, paper and board;

(d) Fibre-dyeing factories;

(e) Cellulose-processing and production installations;

(f) Tannery and leather-dressing factories.

9. *Rubber industry*

Manufacture and treatment of elastomer-based products.

10. *Infrastructure projects*

(a) Industrial-estate development projects;

(b) Urban-development projects;

(c) Ski-lifts and cable-cars;

(d) Construction of roads, harbours, including fishing harbours, and airfields (projects not listed in Annex I);

(e) Canalisation and flood-relief works;

(f) Dams and other installations designed to hold water or store it on a long-term basis;

(g) Tramways, elevated and underground railways, suspended lines or similar lines of a particular type, used exclusively or mainly for passenger transport;

(h) Oil and gas pipeline installations;

(i) Installation of long-distance aqueducts;

(j) Yacht marinas.

11. *Other projects*

(a) Holiday villages, hotel complexes;

(b) Permanent racing and test tracks for cars and motor cycles;

(c) Installations for the disposal of industrial and domestic waste (unless included in Annex I);

(d) Waste water treatment plants;

(e) Sludge-deposition sites;

(f) Storage of scrap iron;

(g) Test benches for engines, turbines or reactors;

(h) Manufacture of artificial mineral fibres;

(i) Manufacture, packing, loading or placing in cartridges of gunpowder and explosives;

(j) Knackers' yards.

12. Modifications to development projects included in Annex I and projects in Annex I undertaken exclusively or mainly for the development and testing of new methods or products and not used for more than one year.

ANNEX III
INFORMATION REFERRED TO IN ARTICLE 5(1)

1. Description of the project, including in particular:

— a description of the physical characteristics of the whole project and the land-use requirements during the construction and operational phases;

— a description of the main characteristics of the production processes, for instance, nature and quantity of the materials used;

— an estimate, by type and quantity, of expected residues and emissions (water, air and soil pollution, noise, vibration, light, heat, radiation, etc) resulting from the operation of the proposed project.

2. Where appropriate, an outline of the main alternatives studied by the developer and an indication of the main reasons for his choice, taking into account the environmental effects.

3. A description of the aspects of the environment likely to be significantly affected by the proposed project, including in particular, population, fauna, flora, soil, water, air, climatic factors, material assets, including the architectural and archaeological heritage, landscape and the inter-relationship between the above factors.

4. A description of the likely significant effects of the proposed project on the environment resulting from:

   — the existence of the project;

   — the use of natural resources;

   — the emission of pollutants, the creation of nuisances and the elimination of waste;

   and the description by the developer of the forecasting methods used to assess the effects on the environment.

5. A description of the measures envisaged to prevent, reduce and where possible offset any significant adverse effects on the environment.

6. A non-technical summary of the information provided under the above headings.

7. An indication of any difficulties (technical deficiencies or lack of know-how) encountered by the developer in compiling the required information.

(Ref: OJ No. L175/40, 5.7.85)

Appendix 2

Schedule 1 and Schedule 2 of The Town and
Country Planning (Assessment of Environmental
Effects) Regulations 1988 (as amended)

SCHEDULE 1
DESCRIPTIONS OF DEVELOPMENT

(1)   The carrying out of building or other operations, or the change of use
        of buildings or other land (where a material change) to provide any
        of the following:

1. A crude-oil refinery (excluding an undertaking manufacturing only
   lubricants from crude oil) or an installation for the gasification and
   liquefication of 500 tonnes or more of coal or bitumous shale per
   day.

2. (a) A thermal power station or other combustion installation
        with a heat output of 300 megawatts or more (not being an
        installation falling within paragraph (b)); and
   (b) A nuclear power station or other nuclear reactor (excluding
        a research installation for the production and conversion of
        fissionable and fertile materials, the maximum power of which
        does not exceed 1 kilowatt continuous thermal load).

3. An installation designed solely for the permanent storage or final
   disposal of radioactive waste.

4. An integrated works for the initial melting of cast-iron and steel.

5. An installation for the extraction of asbestos or for the processing
   and transformation of asbestos or products containing asbestos:

   (a) where the installation produces asbestos-cement products,
        with an annual production of more than 20,000 tonnes of
        finished products; or

   (b) where the installation produces friction material, with an
        annual production of more than 50 tonnes of finished products;
        or

   (c) in other cases, where the installation will utilise more than
        200 tonnes of asbestos per year.

6. An integral chemical installation, that is to say, an industrial installation or group of installations where two or more linked chemical or physical processes are employed for the manufacture of olefins from petroleum products, or of sulphuric acid, nitric acid, hydrofluoric acid, chlorine or fluourine.

7. A special road; a line for long-distance railway traffic; or an aerodrome with a basic runway length of 2,100 metres or more.

8. A trading port, an inland waterway which permits the passage of vessels of over 1,350 tonnes or a port for inland waterway traffic capable of handling such vessels.

9. A waste-disposal installation for the incineration or chemical treatment of special waste.

(2)   The carrying out of operations whereby land is filled with special waste, or the change of use of land (where a material change) to use for the deposit of such waste.

SCHEDULE 2
DESCRIPTIONS OF DEVELOPMENT

Development for any of the following purposes:

1. *Agriculture*
   (a) water-management for agriculture
   (b) poultry-rearing
   (c) pig-rearing
   (d) a salmon hatchery
   (e) an installation for the rearing of salmon
   (f) the reclamation of land from the sea

2. *Extractive industry*
   (a) extracting peat
   (b) deep drilling, including in particular:
       — geothermal drilling
       — drilling for the storage of nuclear waste material
       — drilling for water supplies
       but excluding drilling to investigate the stability of the soil
   (c) extracting minerals (other than metalliferous and energy-producing minerals) such as marble, sand, gravel, shale, salt, phosphates and potash
   (d) extracting coal or lignite by underground or open-cast mining
   (e) extracting petroleum
   (f) extracting natural gas
   (g) extracting ores
   (h) extracting bitumous shale
   (i) extracting minerals (other than metalliferous and energy-producing minerals) by open-cast mining
   (j) a surface industrial installation for the extraction of coal, petroleum, natural gas or ores, or bituminous shale
   (k) a coke oven (dry distillation of coal)
   (l) an installation for the manufacture of cement

3. *Energy industry*
   (a) a non-nuclear thermal power station, not being an installation falling within Schedule 1, or an installation for the production of electricity, steam and hot water
   (b) an industrial installation for carrying gas, steam or hot water; or the transmission of electrical energy by overhead cables
   (c) the surface storage of natural gas
   (d) the underground storage of combustible gases
   (e) the surface storage of fossil fuels

(f)   the industrial briquetting of coal or lignite

(g)   an installation for the production or enrichment of nuclear fuels

(h)   an installation for the reprocessing of irradiated nuclear fuels

(i)   an installation for the collection or processing of radioactive waste, not being an installation falling within Schedule 1

(j)   an installation for hydroelectric energy production

(k)   a wind generator

4.   *Processing of metals*

(a)   an ironworks or steelworks, including a foundry, forge, drawing plant or rolling mill (not being a works falling within Schedule 1)

(b)   an installation for the production (including smelting, refining, drawing and rolling) of non-ferrous metals, other than precious metals

(c)   the pressing, drawing or stamping of large castings

(d)   the surface treatment and coating of metals

(e)   boilermaking or manufacturing reservoirs, tanks and other sheet-metal containers

(f)   manufacturing or assembling motor vehicles or manufacturing motor-vehicle engines

(g)   a shipyard

(h)   an installation for the construction or repair of aircraft

(i)   the manufacture of railway equipment

(j)   swaging by explosives

(k)   an installation for the roasting or sintering of metallic ores

5.   *Glass making*
The manufacture of glass

6.   *Chemical Industry*

(a)   the treatment of intermediate products and production of chemicals, other than development falling within Schedule 1

(b)   the production of pesticides or pharmaceutical products, paints or varnishes, elastomers or peroxides

(c)   the storage of petroleum or petrochemical or chemical products

7.   *Food Industry*

(a)   the manufacture of vegetable or animal oils or fats

(b)   the packing or canning of animal or vegetable products

(c)   the manufacture of dairy products

(d)   brewing or malting

(e)   confectionery or syrup manufacture

(f)   an installation for the slaughter of animals

(g)   an industrial starch manufacturing installation

(h) a fish-meal or fish-oil factory

(i) a sugar factory

8. *Textile, leather, wood and paper industries*
    (a) a wood scouring, degreasing and bleaching factory
    (b) the manufacture of fibre board, particle board or plywood
    (c) the manufacture of pulp, paper or board
    (d) a fibre-dyeing factory
    (e) a cellulose-processing and production installation
    (f) a tannery or a leather dressing factory

9. *Rubber Industry*
    The manufacture and treatment of elastomer-based products

10. *Infrastructure projects*
    (a) an industrial estate development project
    (b) an urban development project
    (c) a ski-lift or cable-car
    (d) the construction of a road, or a harbour, including a fishing harbour, or an aerodrome, not being development falling within Schedule 1
    (e) canalisation or flood-relief works
    (f) a dam or other installation designed to hold water or store it on a long-term basis
    (g) a tramway, elevated or underground railway, suspended line or similar line, exclusively or mainly for passenger transport
    (h) an oil or gas pipeline installation
    (i) a long-distance aqueduct
    (j) a yacht marina
    (k) a motorway service area
    (l) coast protection works

11. *Other projects*
    (a) a holiday village or hotel complex
    (b) a permanent racing or test track for cars or motor cycles
    (c) an installation for the disposal of controlled waste from mines and quarries, not being an installation falling within Schedule 1
    (d) a waste-water treatment plant
    (e) a site for depositing sludge
    (f) the storage of scrap iron
    (g) a test bench for engines, turbines or reactors
    (h) the manufacture of artificial mineral fibres
    (i) the manufacture, packing, loading or placing in cartridges of gunpowder or other explosives
    (j) a knacker's yard

12. The modification of a development which has been carried out, where that development is within a description mentioned in Schedule 1.

13. Development within a description mentioned in Schedule 1, where it is exclusively or mainly for the development and testing of new methods or products and will not be permitted for longer than one year.

(Ref: SI 1990/367 as amended by SI 1992/1492 and SI 1994/677)

Appendix 3

Schedule 3 of The Town and Country Planning (Assessment of Environmental Effects) Regulations 1988 (as amended)

SCHEDULE 3

1. An environmental statement comprises a document or series of documents providing for the purpose of assessing the likely impact upon the environment of the development proposed to be carried out, the information specified in paragraph 2 (referred to in this Schedule as "the specified information").

2. The specified information is:
   - (a) a description of the development proposed, comprising information about the site and the design and size or scale of the development;
   - (b) the data necessary to identify and assess the main effects which that development is likely to have on the environment;
   - (c) a description of the likely significant effects, direct and indirect, on the environment of the development, explained by reference to its possible impact on:

     human beings;
     flora;
     fauna;
     soil;
     water;
     air;
     climate;
     the landscape;
     the inter-action between any of the foregoing;
     material assets;
     the cultural heritage;

   - (d) where significant adverse effects are identified with respect to any of the foregoing, a description of the measures envisaged in order to avoid, reduce or remedy those effects; and
   - (e) a summary in non-technical language of the information specified above.

3. An environmental statement may include, by way of explanation or amplification of any specified information, further information on any of the following matters:

(a) the physical characteristics of the proposed development, and the land-use requirements during the construction and operational phases;

(b) the main characteristics of the production processes proposed, including the nature and quality of the materials to be used;

(c) the estimated type and quantity of expected residues and emissions (including pollutants of water, air or soil, noise, vibration, light, heat and radiation) resulting from the proposed development when in operation;

(d) (in outline) the main alternatives (if any) studied by the applicant, appellant or authority and an indication of the main reasons for choosing the development proposed, taking into account the environmental effects;

(e) the likely significant direct and indirect effects on the environment of the development proposed which may result from:

(i)   the use of natural resources;

(ii) the emission of pollutants, the creation of nuisances, and the elimination of waste;

(f) the forecasting methods used to assess any effects on the environment about which information is given under sub-paragraph (e); and

(g) any difficulties, such as technical deficiencies or lack of know-how, encountered in compiling any specified information.

In sub-paragraph (e), "effects" includes secondary, cumulative, short, medium and long term, permanent, temporary, positive and negative effects.

Where further information is included in an environmental statement pursuant to paragraph 3, a non-technical summary of that information shall also be provided.

(Ref: SI 1990/367 as amended by SI 1992/1494 and SI 1994/677)

Appendix 4

The text of Council Directive 97/11/EC of 3 March 1997 amending Directive 85/337/EEC on the assessment of the effects of certain public and private projects on the environment

THE COUNCIL OF THE EUROPEAN UNION,

Having regard to the Treaty establishing the European Community, and in particular Article 130s (1) thereof,

Having regard to the proposal from the Commission,

Having regard to the opinion of the Economic and Social Committee,

Having regard to the opinion of the Committee of the Regions,

Acting in accordance with the procedure laid down in Article 189c of the Treaty,

(1)   Whereas Council Directive 85/337/EEC of 27 June 1985 on the assessment of the effects of certain public and private projects on the environment aims at providing the competent authorities with relevant information to enable them to take a decision on a specific project in full knowledge of the project's likely significant impact on the environment; whereas the assessment procedure is a fundamental instrument of environmental policy as defined in Article 130r of the Treaty and of the Fifth Community Programme of policy and action in relation to the environment and sustainable development;

(2)   Whereas, pursuant to Article 130r (2) of the Treaty, Community policy on the environment is based on the precautionary principle and on the principle that preventative action should be taken, that environmental damage should as a priority be rectified at source and that the polluter should pay;

(3)   Whereas the main principles of the assessment of environmental effects should be harmonised and whereas the Member States may lay down stricter rules to protect the environment;

(4)   Whereas experience acquired in environmental impact assessment, as recorded in the report on the implementation of Directive 85/337/EEC, adopted by the Commission on 2 April 1993, shows that

it is necessary to introduce provisions designed to clarify, supplement and improve the rules on the assessment procedure, in order to ensure that the Directive is applied in an increasingly harmonised and efficient manner;

(5)   Whereas projects for which an assessment is required should be subject to a requirement for development consent; whereas the assessment should be carried out before such consent is granted;

(6)   Whereas it is appropriate to make additions to the list of projects which have significant effects on the environment and which must on that account as a rule be made subject to systematic assessment;

(7)   Whereas projects of other types may not have significant effects on the environment in every case; whereas these projects should be assessed where Member States consider they are likely to have significant effects on the environment;

(8)   Whereas Member States may set thresholds or criteria for the purpose of determining which such projects should be subject to assessment on the basis of the significance of their environmental effects; whereas Member States should not be required to examine projects below those thresholds or outside those criteria on a case-by-case basis;

(9)   Whereas when setting such thresholds or criteria or examining projects on a case-by-case basis for the purpose of determining which projects should be subject to assessment on the basis of their significant environmental effects, member States should take account of the relevant selection criteria set out in this Directive; whereas, in accordance with the subsidiary principle, the Member States are in the best position to apply these criteria in specific instances;

(10)   Whereas the existence of a location criterion referring to special protection areas designated by Member States pursuant to Council Directive 79/409/EEC of 2 April 1979 on the conservation of wild birds and 92/43/EEC of 21 May 1992 on the conservation of natural habitats and of wild fauna and flora does not imply necessarily that projects in those areas are to be automatically subject to an assessment under this Directive;

(11)   Whereas it is appropriate to introduce a procedure in order to enable the developer to obtain an opinion from the competent authorities on the content and extent of the information to be elaborated and supplied for the assessment; whereas Member States' in the framework of this

procedure, may require the developer to provide, *inter alia*, alternatives for the projects for which it intends to submit an application;

(12)   Whereas it is desirable to strengthen the provisions concerning environmental impact assessment in a transboundary context to take account of developments at international level;

(13)   Whereas the Community signed the Convention on Environmental Impact Assessment in a Transboundary Context on 25 February 1991,

HAS ADOPTED THIS DIRECTIVE:

Article 1

Directive 85/337/EEC is hereby amended as follows:

1.   Article 2(1) shall be replaced by the following:
      "1.   Member States shall adopt all measures necessary to ensure that, before consent is given, projects likely to have significant effects on the environment by virtue, *inter alia*, of their nature, size or location are made subject to a requirement for development consent and an assessment with regard to their effects. These projects are defined in Article 4."

      2.   The following paragraph shall be inserted in Article 2:
2.   "2a.   Member States may provide for a single procedure in order to fulfil the requirements of this Directive and the requirements of Council Directive 96/61/EC of 24 September 1996 on integrated pollution prevention and control."

3.   The first subparagraph of Article 2(3) shall read as follows:
      "3.   Without prejudice to Article 7, Member States may, in exceptional cases, exempt a specific project in whole or in part from the provisions laid down in this Directive."

4.   In Article 2(3)(c) the words "where appropriate" shall be replaced by the words "where applicable";

5.   Article 3 shall be replaced by the following:

"*Article 3*

The environmental impact assessment shall identify, describe and assess in an appropriate manner, in the light of each individual case and in accordance with Articles 4 to 11, the direct and indirect effects of a project on the following factors:

— human beings, fauna and flora;
— soil, water, air, climate and the landscape;
— material assets and the cultural heritage;
— the interaction between the factors mentioned in the first, second and third indents."

6. Article 4 shall be replaced by the following:

"*Article 4*

1. Subject to Article 2(3), projects listed in Annex I shall be made subject to an assessment in accordance with Articles 5 to 10.
2. Subject to Article 2(3), for projects listed in Annex II, the Member States shall determine through:

(a) a case-by-case examination; or
(b) thresholds or criteria set by the Member State

whether the project shall be made subject to an assessment in accordance with Articles 5 to 10.
Member States may decide to apply both procedures referred to in (a) and (b).

3. When a case-by-case examination is carried out or thresholds or criteria are set for the purpose of paragraph 2, the relevant selection criteria set out in Annex III shall be taken into account.
4. Member States shall ensure that the determination made by the competent authorities under paragraph 2 is made available to the public."

7. Article 5 shall be replaced by the following:

"*Article 5*

1. In the case of projects which, pursuant to Article 4, must be subjected to an environmental impact assessment in accordance with Articles 5 to 10, Member States shall adopt the necessary measures to ensure that the developer supplies in an appropriate form the information specified in Annex IV inasmuch as:

(a) the Member States consider that the information is relevant to a given stage of the consent procedure and to the specific characteristics of a particular product or type of project and of the environmental features likely to be affected;
(b) the Member States consider that a developer may reasonably be

required to compile this information having regard *inter alia* to current knowledge and methods of assessment.

2. Member States shall take the necessary measures to ensure that, if the developer so requests before submitting an application for development consent, the competent authority shall give an opinion on the information to be supplied by the developer in accordance with paragraph 1. The competent authority shall consult the developer and authorities referred to in Article 6(1) before it gives its opinion. The fact that the authority has given an opinion under this paragraph shall not preclude it from subsequently requiring the developer to submit further information.

Member States may require the competent authorities to give such an opinion, irrespective of whether the developer so requests.

3. The information to be provided by the developer in accordance with paragraph 1 shall include at least:

— a description of the project comprising information on the site, design and size of the project,

— a description of the measures envisaged in order to avoid, reduce and, if possible, remedy significant adverse effects,

— the data required to identify and assess the main effects which the project is likely to have on the environment,

— an outline of the main alternatives studied by the developer and an indication of the main reasons for his choice, taking into account the environmental effects,

— a non-technical summary of the information mentioned in the previous indents.

4. Member States shall, if necessary, ensure that any authorities holding relevant information, with particular reference to Article 3, shall make this information available to the developer."

8. Article 6(1) shall be replaced by the following:

"1. Member States shall take the measures necessary to ensure that the authorities likely to be concerned by the project by reason of their specific environmental responsibilities are given an opportunity to express their opinion on the information supplied by the developer and on the request for development consent. To this end, Member States shall designate the authorities to be consulted, either in general terms or on a case-by-case basis. The information gathered pursuant to Article 5 shall be forwarded to those authorities. Detailed arrangements for consultation shall be laid down by the Member States."

Article 6(2) shall be replaced by the following:

"2. Member States shall ensure that any request for development consent and any information gathered pursuant to Article 5 are made available to the public within a reasonable time in order to give the public concerned the opportunity to express an opinion before the development consent is granted."

9.   Article 7 shall be replaced by the following:

"*Article 7*

1.   Where a Member state is aware that a project is likely to have significant effects on the environment in another Member State or where a Member State likely to be significantly affected so requests, the Member State in whose territory the project is intended to be carried out shall send to the affected Member State as soon as possible and no later than when informing its own public, *inter alia*:

(a)   a description of the project, together with any available information on its possible transboundary impact;

(b)   information on the nature of the decision which may be taken,

and shall give the other Member State a reasonable time in which to indicate whether it wishes to participate in the Environmental Impact Assessment procedure, and may include the information referred to in paragraph 2.

2.   If a Member State which receives information pursuant to paragraph 1 indicates that it intends to participate in the Environmental Impact Assessment procedure, the Member State in whose territory the project is intended to be carried out shall, if it has not already done so, send to the affected Member State the information gathered pursuant to Article 5 and relevant information regarding the said procedure, including the request for development consent.

3.   The Member States concerned, each insofar as it is concerned, shall also:

(a)   arrange for the information referred to in paragraphs 1 and 2 to be made available, within a reasonable time, to the authorities referred to in Article 6(1) and the public concerned in the territory of the Member State likely to be significantly affected; and

(b)   ensure that those authorities and the public concerned are given an opportunity, before development consent for the project is granted, to forward their opinion within a reasonable time on the information supplied to the competent authority in the Member State in whose territory the project is intended to be carried out.

4. The Member States concerned shall enter into consultations regarding, *inter alia*, the potential transboundary effects of the project and the measures envisaged to reduce or eliminate such effects and shall agree on a reasonable time frame for the duration of the consultation period.

5. The detailed arrangements for implementing the provisions of this Article may be determined by the Member States concerned."

10. Article 8 shall be replaced by the following:

"*Article 8*

The results of consultations and the information gathered pursuant to Articles 5, 6, and 7 must be taken into consideration in the development consent procedure."

11. Article 9 shall be replaced by the following:

"*Article 9*

1. When a decision to grant or refuse development consent has been taken, the competent authority or authorities shall inform the public thereof in accordance with the appropriate procedures and shall make available to the public the following information:
   — the content of the decision and any conditions attached thereto,
   — the main reasons and considerations on which the decision is based,
   — a description, where necessary, of the main measures to avoid, reduce and, if possible, offset the major adverse effects.

2. The competent authority or authorities shall inform any Member State which has been consulted pursuant to Article 7, forwarding to it the information referred to in paragraph 1."

12. Article 10 shall be replaced by the following:

"*Article 10*

The provisions of this Directive shall not affect the obligation on the competent authorities to respect the limitations imposed by national regulations and administrative provisions and accepted legal practices with regard to commercial and industrial confidentiality, including intellectual property, and the safeguarding of the public interest.

Where Article 7 applies, the transmission of information to another Member State and the receipt of information by another Member State shall be subject to the limitations in force in the Member State in which the project is proposed."

13. Article 11(2) shall be replaced by the following:
    "2.   In particular, Member States shall inform the Commission of any criteria and/or thresholds adopted for the selection of the projects in question, in accordance with Article 4(2)."

14. Article 13 shall be deleted.

15. Annexes I, II and III shall be replaced by Annexes I, II, III and IV as they appear in the Annex.

Article 2

Five years after the entry into force of this Directive, the Commission shall send the European Parliament and Council a report on the application and effectiveness Directive 85/337/EEC as amended by this Directive. Its report shall be based on the exchange of information provided for by Article 11(1) and (2).

On the basis of this report, the Commission shall, where appropriate, submit to the Council additional proposals with a view to ensuring further co-ordination in the application of this Directive.

Article 3

Member States shall bring into force the laws, regulations and administrative provisions necessary to comply with this Directive by 14 March 1999 at the latest. They shall forthwith inform the Commission thereof.

When Member States adopt these provisions, they shall contain a reference to this Directive or shall be accompanied by such reference at the time of their official publication. The procedure for such reference shall be adopted by Member States.

If a request for development consent is submitted to a competent authority before the end of the time limit laid down in paragraph 1, the provisions of Directive 85/337/EEC prior to these amendments shall continue to apply.

Article 4

This Directive shall enter into force on the twentieth day following that of its publication in the *Official Journal of the European Communities*.

Article 5

This Directive is addressed to the Member States.

Done at Brussels, 3 March 1997.

For the Council

The President

M. DE BOER

ANNEX 1

PROJECTS SUBJECT TO ARTICLE 4(1)

1. Crude-oil refineries (excluding undertakings manufacturing only lubricants from crude oil) and installations for the gasification and liquefication of 500 tonnes or more of coal or bituminous shale per day.

2. (i)   Thermal power stations and other combustion installations with a heat output of 300 megawatts or more, and
    (ii) nuclear power stations and other nuclear reactors including the dismantling or decommissioning of such power stations or reactors (except research installations for the production and conversion of fissionable and fertile materials, whose maximum power does not exceed 1 kilowatt continuous thermal load).

3. (a)   Installations for the reprocessing of irradiated nuclear fuel.
    (b)   Installations designed:
          (i)   for the production or enrichment of nuclear fuel,
          (ii) for the processing of irradiated nuclear fuel or high-level radioactive waste,
          (iii) for the final disposal of radioactive waste,
          (iv) solely for the storage (planned for more than 10 years) of irradiated nuclear fuels or radioactive waste in a different site than the production site.

4. (i)   Integrated works for the initial smelting of cast-iron and steel;
    (ii) Installations for the production of non-ferrous crude metals from ore, concentrates or secondary raw materials by metallurgical, chemical or electrolytic processes.

5. Installations for the extraction of asbestos and for the processing and transformation of asbestos and products containing asbestos: for asbestos-cement products, with an annual production of more than 20,000 tonnes of finished products, for friction material, with an annual production of more than 50 tonnes of finished products, and for other uses of asbestos, utilisation of more than 200 tonnes per year.

6. Integrated chemical installations, i.e. those installations for the manufacture on an industrial scale of substances using chemical conversion processes, in which several units are juxtaposed and are functionally lined to one another and which are:
    (a)   for the production of basic organic chemicals;
    (b)   for the production of basic inorganic chemicals;

(c)   for the production of phosphorous-, nitrogen-, or potassium-based fertilisers (simple or compound fertilisers);

(d)   for the production of basic plant health products and of biocides;

(e)   for the production of basic pharmaceutical products using a chemical or biological process;

(f)   for the production of explosives.

7.   (a)   Construction of lines for long-distance railway traffic and of airports with a basic runway length of 2,100 m or more;

(b)   Construction of motorways and express roads;

(c)   Construction of a new road of four or more lanes, or realignment and/or widening of an existing road of two lanes or less so as to provide four or more lanes, where such new road, or realigned and/or widened section of road would be 10km or more in a continuous length.

8.   (a)   Inland waterways and ports for inland-waterway traffic which permit the passage of vessels of over 1,350 tonnes;

(b)   Trading ports, piers for loading and unloading connected to land and outside ports (excluding ferry piers) which can take vessels of over 1,350 tonnes.

9.   Waste disposal installations for the incineration, chemical treatment as defined in Annex IIA to Directive 75/442/EEC under heading D9, or landfill of hazardous waste (i.e. waste to which Directive 91/689/EEC applies).

10.   Waste disposal installations for the incineration or chemical treatment as defined in Annex IIA to Directive 75/442/EEC under heading D9 of non-hazardous waste with a capacity exceeding 100 tonnes per day.

11.   Groundwater abstraction or artificial groundwater recharge schemes where the annual volume of water abstracted or recharged is equivalent to or exceeds 10 million cubic metres.

12.   (a)   Works for the transfer of water resources between river basins where this transfer aims at preventing possible shortages of water and where the amount of water transferred exceeds 100 million cubic metres/year;

(b)   In all other cases, works for the transfer of water resources between river basins where the multi-annual average flow of the basin of abstraction exceeds 2,000 million cubic metres/year and where the amount of water transferred exceeds 5% of this flow.

In both cases transfers of piped drinking water are excluded.

253

13. Waste water treatment plants with a capacity exceeding 150,000 population equivalent as defined in Article 2 point (6) of Directive 91/271.EEC.

14. Extraction of petroleum and natural gas for commercial purposes where the amount extracted exceeds 500 tonnes/day in the case of petroleum and 500,000 m^3/day in the case of gas.

15. Dams and other installations designed for the holding back or permanent storage of water, where a new or additional amount of water held back or stored exceeds 10 million cubic metres.

16. Pipelines for the transport of gas, oil or chemicals with a diameter of more than 800 mm and a length of more than 40 km.

17. Installations for the intensive rearing of poultry or pigs with more than:
    (a)   85,000 places for broilers, 60,000 placers for hens;
    (b)   3,000 places for production pigs (over 30 kg); or
    (c)   900 places for sows.

18. Industrial plants for the:
    (a)   production of pulp from timber or similar fibrous materials;
    (b)   production of paper and board with a production capacity exceeding 200 tonnes per day.

19. Quarries and open-cast mining where the surface of the site exceeds 25 hectares, or peat extraction, where the surface of the site exceeds 150 hectares.

20. Construction of overhead electrical power lines with a voltage of 220 kV or more and a length of more than 15 km.

21. Installations for storage of petroleum, petrochemical, or chemical products with a capacity of 200,000 tonnes or more.

ANNEX II

PROJECTS SUBJECT TO ARTICLE 4(2)

1. Agriculture, silviculture and aquaculture:
    (a)   Projects for the restructuring of rural land holdings;
    (b)   Projects for the use of uncultivated land or semi-natural areas for intensive agricultural purposes;
    (c)   Water management projects for agriculture, including irrigation and land drainage projects;
    (d)   Initial afforestation and deforestation for the purposes of conversion to another type of land use;
    (e)   Intensive livestock installations (projects not included in Annex I);
    (f)   Intensive fish farming;
    (g)   Reclamation of land from the sea.

2. Extractive industry
    (a)   Quarries, open-cast mining and peat extraction (projects not included in Annex I);
    (b)   Underground mining;
    (c)   Extraction of minerals by marine or fluvial dredging;
    (d)   Deep drillings, in particular:
        (i) geothermal drilling,
        (ii) drilling for the storage of nuclear waste material,
        (iii) drilling for water supplies,
        with the exception of drillings for investigating the stability of the soil;
    (e)   Surface industrial installations for the extraction of coal, petroleum, natural gas and ores, as well as bituminous shale.

3. Energy industry
    (a)   Industrial installations for the production of electricity, steam and hot water (projects not included in Annex I);
    (b)   Industrial installations for carrying gas, steam and hot water, transmission of electrical energy by overhead cables (projects not included in Annex I);
    (c)   Surface storage of natural gas;
    (d)   Underground storage of combustible gases;
    (e)   Surface storage of fossil fuels;
    (f)   Industrial briquetting of coal and lignite;
    (g)   Installations for the processing and storage of radioactive waste (unless included in Annex I);
    (h)   Installations for hydroelectric energy production;

(i)   Installations for the harnessing of wind power for energy production (wind farms).

4. Production and processing of metals
   (a)   Installations for the production of pig iron or steel (primary or secondary fusion) including continuous casting;
   (b)   Installations for the processing of ferrous metals:
          (i)   hot-rolling mills;
          (ii) smitheries with hammers;
          (iii) application of protective fused metal coats;
   (c)   Ferrous metal foundries;
   (d)   Installations for the smelting, including the alloyage, of non-ferrous metals, excluding precious metals, including recovered products (refining, foundry casting, etc.);
   (e)   Installations for surface treatment of metals and plastic materials using an electrolytic or chemical process;
   (f)   Manufacture and assembly of motor vehicles and manufacture of motor-vehicle engines;
   (g)   Shipyards;
   (h)   Installations for the construction and repair of aircraft;
   (i)   Manufacture of railway equipment;
   (j)   Swaging by explosives;
   (k)   Installations for the roasting and sintering of metallic ores.

5. Mineral industry
   (a)   Coke ovens (dry coal distillation);
   (b)   Installations for the manufacture of cement;
   (c)   Installations for the production of asbestos and the manufacture of asbestos-products (projects not included in Annex I);
   (d)   Installations for the manufacture of glass including glass fibre;
   (e)   Installations for smelting mineral substances including the production of mineral fibres;
   (f)   Manufacture of ceramic products by burning, in particular roofing tiles, bricks, refractory bricks, tiles, stoneware or porcelain.

6. Chemical industry (projects not included in Annex I)
   (a)   Treatment of intermediate products and production of chemicals;
   (b)   Production of pesticides and pharmaceutical products, paint and varnishes, elastomers and peroxides;
   (c)   Storage facilities for petroleum, petrochemical and chemical products.

7. Food industry
   (a)   Manufacture of vegetable and animal oils and fats;

(b)   Packaging and canning of animal and vegetable products;

(c)   Manufacture of dairy products;

(d)   Brewing and malting;

(e)   Confectionery and syrup manufacture;

(f)   Installations for the slaughter of animals;

(g)   Industrial starch manufacturing installations;

(h)   Fish-meal and fish-oil factories;

(i)   Sugar factories.

8. Textile, leather, wood and paper industries

(a)   Industrial plants for the production of paper and board (projects not included in Annex I);

(b)   Plants for the pretreatment (operations such as washing, bleaching, mercerization) or dyeing of fibres or textiles;

(c)   Plants for the tanning of hides and skins;

(d)   Cellulose-processing and production installations.

9. Rubber industry

Manufacture and treatment of elastomer-based products.

10. Infrastructure projects

(a)   Industrial estate development projects;

(b)   Urban development projects, including the construction of shopping centres and car parks;

(c)   Construction of railways and intermodal transshipment facilities and of intermodal terminals (projects not included in Annex I);

(d)   Construction of airfields (projects not included in Annex I);

(e)   Construction of roads, harbours and port installations, including fishing harbours (projects not included in Annex I);

(f)   Inland-waterway construction not included as Annex I, canalization and flood-relief works;

(g)   Dams and other installations designed to hold water or store it on a long-term basis (projects not included in Annex I);

(h)   Tramways, elevated and underground railways, suspended lines or similar lines of a particular type, used exclusively or mainly for passenger transport;

(i)   Oil and gas pipeline installations (projects not included in Annex I);

(j)   Installations of long-distance aqueducts;

(k)   Coastal work to combat erosion and maritime works capable of altering the coast through the construction, for example, of dykes, moles, jetties and other sea defence works, excluding the maintenance and reconstruction of such works;

(l)   Groundwater abstraction and artificial groundwater recharge schemes not included in Annex I;

(m)   Works for the transfer of water resources between river basins not included in Annex I.

11. Other projects

(a)   Permanent racing and test tracks for motorized vehicles;

(b)   Installations for the disposal of waste (projects not included in Annex I);

(c)   Waste-water treatment plants (projects not included in Annex I);

(d)   Sludge-deposition sites;

(e)   Storage of scrap iron, including scrap vehicles;

(f)   Test benches for engines, turbines or reactors;

(g)   Installations for the manufacture of artificial mineral fibres;

(h)   Installations for the recovery or destruction of explosive substances;

(i)   Knacker's yards.

12. Tourism and leisure

(a)   Ski-runs, ski-lifts and cable-cars and associated developments;

(b)   Marinas;

(c)   Holiday villages and hotel complexes outside urban areas and associated developments;

(d)   Permanent camp sites and caravan sites;

(e)   Theme parks.

13. Any change or extension of projects listed in Annex I or Annex II, already authorized, executed or in the process of being executed, which may have significant adverse effects on the environment;

Projects in Annex I, undertaken exclusively or mainly for the development and testing of new methods or products and not used for more than two years.

ANNEX III

SELECTION CRITERIA REFERRED TO IN ARTICLE 4(3)

1. Characteristics of projects
   — the size of the project,
   — the cumulation with other projects,
   — the use of natural resources,
   — the production of waste
   — pollution and nuisances
   — the risk of accidents, having regard in particular to substances or technologies used.

2. Location of projects

   The environmental sensitivity of geographical areas likely to be affected by projects must be considered, having regard, in particular, to:
   — the existing land use,
   — the relative abundance, quality and regenerative capacity of natural resources in the area,
   — the absorption capacity of the natural environment, paying particular attention to the following areas:
   (a) wetlands
   (b) coastal zones;
   (c) mountain and forest areas;
   (d) nature reserves and parks;
   (e) areas classified or protected under Member States' legislation; special protection areas designated by Member States pursuant to Directive 79/409/EEC and 92/43/EEC;
   (f) areas in which the environmental quality standards laid down in Community legislation have already been exceeded;
   (g) densely populated areas;
   (h) landscapes of historical, cultural or archaeological significance.

3. Characteristics of the potential impact

   The potential significant effects of projects must be considered in relation to criteria set out under 1 and 2 above, and having regard to:
   — the extent of the impact (geographical area and size of the affected population),
   — the transfrontier nature of the impact,
   — the magnitude and complexity of the impact,
   — the probability of the impact,
   — the duration, frequency and reversibility of the impact.

ANNEX IV

INFORMATION REFERRED TO IN ARTICLE 5(1)

1. Description of the projects, including in particular:
   — a description of the physical characteristics of the whole project and the land-use requirements during the construction and operational phases.
   — a description of the main characteristics of the production processes, for instance, nature and quantity of the materials used.
   — an estimate, by type and quantity, of expected residues and emissions (water, air and soil pollution, noise, vibration, light, heat, radiation, etc.) resulting from the operation of the proposed project.

2. An outline of the main alternative studied by the developer and an indication of the main reasons for this choice, taking into account the environmental effects.

3. A description of the aspects of the environment likely to be significantly affected by the proposed project, including, in particular, population, fauna, flora, soil, water, air, climatic factors, material assets, including the architectural and archaeological heritage, landscape and the inter-relationship between the above factors.

4. A description of the likely significant effects of the proposed project on the environment resulting from:
   — the existence of the project,
   — the use of natural resources,
   — the emission of pollutants, the creation of nuisances and the elimination of waste,
   and the description by the developer of the forecasting methods used to assess the effects on the environment.

5. A description of the measures envisaged to prevent, reduce and where possible offset any significant adverse effects on the environment.

6. A non-technical summary of the information provided under the above headings.

7. An indication of any difficulties (technical deficiencies or lack of know-how) encountered by the developer in compiling the required information.

(Ref: OJ No. L73/13, 14.03.97)

Appendix 5
The Internet and environmental assessment

Introduction

The almost exponential growth of the World Wide Web (www) has turned the Internet into possibly the main global communication system and information supplier. The Internet offers a wide range of utilities (information, databases, contacts, business opportunities, communication etc.) to most facets of society (public bodies, lobby groups, non-governmental organisations, researchers, professionals, private companies, the public in general, etc.) and that the www is expected to become a key tool for many professionals in their daily work; through the vast growth in the number of individuals and companies becoming registered with Internet access it is apparent that this is already occurring. Estimates put the number of people accessing (or surfing) the Internet each day between 30 and 50 million, rising by 2·3% each week, while the number of computers connected to the Internet full-time stood at 13 million in July 1996 (survey by Network Wizards at http://www.nw.com.)

This Appendix presents the results and conclusions of research undertaken during 1997 to evaluate and assess the usefulness of the Internet to the environmental impact assessment process. As the Internet is growing at a exceptional rate, with web sites closing down, being amended and being added all the time, any research of this kind quickly becomes out of date, but we considered the information to be valuable for inclusion within this book.

The research had two main aims:

- firstly, to identify what role the Internet could (and does) play in the environmental assessment process; and
- secondly, to identify short cuts and useful sites for users of this book.

A further aim was to identify potentially useful web sites for inclusion in this book.

It was considered at the outset of the research that the potential roles in which the Internet could best assist were:

- the supply and dissemination of environmental assessment techniques being adopted for the undertaking of an environmental assessment; and

- the carriage and display of information/data relating to a specific environmental assessment being undertaken for a particular type of development.

As identified in earlier chapters, the EIA process involves a number of phases, the completion of which are based on the following elements:

- baseline studies concerning both the proposal and the site;
- consultation and public participation;
- impact identification and assessment methodologies; and
- design of mitigation measures and audit monitoring.

At the beginning of the research it was considered that the greatest potential use of the www would be as a source of information and data relating to specific areas of information (e.g. air quality in Tyneside, or river water quality in the River Severn) as well as a fast and interactive means for public consultation, comment and participation. It was also considered that research on differing or novel approaches to the assessment process, or those adopted throughout the world, may be present on web sites operated by research institutions or lending agencies, and that the Internet would present an alternative medium for researchers to the normal sources of information and data such as conferences, journals and periodicals.

Approach

In order to access a web site it is necessary to know its address. This typically takes the following form:

http://www.who.org.uk/index.htm

where:	
http://	signifies the hypertext transfer protocol
www	indicates the site is on the World Wide Web
who.org.uk	is the domain name of the web site — 'who' is the subscriber's specific abbreviation and the rest of the name depends on the nature of the organisation and its location.
index	is the name of the document
htm	is the format (Hypertext Markup Language)

Domain names in the UK are usually in the form:

co.uk	commercial organisations
org.uk	non-profit organisations
sch.uk	schools
ac.uk	academic institutions, e.g. universities and colleges
plc.uk	public limited companies

ltd.uk   private limited companies

International domain names are usually of the form:

com.xx   commercial organisations
org.xx   non-profit organisations
edu.xx   educational establishments
where 'xx' is a suffix depending on the country (fr, France; jp, Japan, etc.).

In the event that a web-site address is unknown or the aim is to find a number of sites dealing with a specific topic, it is possible to search (or surf) the Internet looking for such sites. To undertake this a 'search engine' is required.

Examples of some engines are given in Table 19. There are many such engines available; they vary in their user-friendliness, efficiency, speed, size and usefulness, but they are indispensable for surfing the net and identifying relevant web sites.

As part of this exercise it was decided to focus on two search engines, Yahoo and Altavista. These were selected primarily because of their technical superiority in method of approach combined with substantial commercial success.

Finding available information on the Net is not straightforward, however, because of the huge number of related and unrelated web sites that may be identified with an incorrect or unspecific search (for example a search for 'site

Table 19. Search engines

Engine name	Web-site address[a]	Comments
Lycos	www.lycos.com	Easy to use, fast, large coverage, free
Yahoo	www.yahoo.com	Good for general searching
Infoseek	www.infoseek.com	Small coverage of sites but very easy to use and therefore efficient, free
Altavista	www.altavista.digital.com	Fast (but very popular, which can result in slowed search speeds), extensive and up-to-date database
Magellan	www.mckinley.com	Good appearance and reliable
Nlightn	www.nlightn.com	Largest database of all search engines but can be slow as a result
Excite	www.excite.com	Large database and relatively quick
Webcrawler	www.webcrawler.com	Good for fast simple searches

[a] All the web-site addresses are prefixed with: 'http://'

investigations' revealed several thousand web sites — far too many to review realistically).

In order to compare the results obtained from these two search engines, and indeed to assess the usefulness of the www and its relevance to the environmental assessment process, a number of keywords were selected. Examples are listed below.

Keywords:

Environment(al) assessment (statement)
Environment(al) database
Environment(al) impact assessment (statement)
Environmental effects
Environmental quality
Environmental impact procedure
Environmental assessment process
Program impact assessment
Regional environmental plans
Environmental Action Plans (EAPs)
Environmental Management Plans (EMPs)
Environmental Response Plans (ERPs)
Environmental implications
Public participation
Scoping
Audit monitoring
Leopoldo matrix
Natural resource effects
Physical effects
Social effects
Existing environment
Air quality impact assessment
Ecological impact assessment
Economic impact assessment
Environmental health impact assessment
Hazard and risk impact assessment
Landscape impact assessment
Noise impact assessment
Social impact assessment
Traffic impact assessment
Public enquiries

Monitoring
Audit monitoring programmes
Pollution control
Urban impact assessment
Visual impact assessment
Water quality impact assessment
Environmental costs
Social and environmental costs and benefits
Direct costs
Indirect costs
Cost–benefit analysis (CBA)
Cost–effectiveness analysis (CEA)
Impact significance
Mitigation measures
Corrective measures
Remedial measurements
Public mediation
Environmental audits
Environmental interaction matrix
Post-project analysis (PPA), auditing
Types of projects:
Agriculture
Business park
Community centre
Pipeline
Expressway
Fish farm
Flood defence
Forestry
Gas processing plant
Gas terminal
Harbour
Hydroelectric power plant
Incinerator
Industrial

264

Landfill
Leisure
Marina
Mineral extraction
Nuclear Power Plant
Open-cut mining operation
Pipeline
Pipeline
Power Station

Residential
Retail
Road
Sand and gravel extraction
operation
Sewage treatment plant
Thermal electricity generating plant
Transmission line
Transport

Yahoo surfing

The way Yahoo is built directly affects the amount and format of the results obtained. In Yahoo the information has already been vetted and categorised according to subject matter by human operators (some search engines employ artificial intelligence which can result in erroneous results). Therefore, the results of searching for a keyword are those available from Yahoo and not from the whole Web. Frequently one obtains a small number of sites; this more readily facilitates the task of reviewing their content. The taxonomy of the listing allows one to focus the search on a particular interest (for example, you could ignore all the sites under the headline 'Business and economy'). Although the categorisation and listing of main sites can save hours of searching, the results can at times be very general and incomprehensive.

Asked to search for the keyword 'environmental impact assessment', Yahoo located 10 sites, of which eight corresponded to environmental consultants offering their services and two belonged to the Environment Protection Division of the Northern Territory Department (Australian Government).

Asked to search for the keyword 'environmental assessment', 133 sites were identified, 85% of which were consultants; the remainder were related to specific research projects, universities and government departments. Most of the latter sites did however have potentially interesting links to other sites. It is noteworthy that a search for 'environment assessment' (as opposed to 'environmental assessment') yielded 157 sites, some of which were different from the ones identified previously.

For 'environmental database' Yahoo located 34 sites. Some of them were commercial web pages offering their products and also providing some free-access resources. There are free-access databases at both government department and non-profit organisation web sites (none of them British).

The EnviroNet Australia (http://www.erin.gov.au/net/environet.html) web site proved very interesting: the home page has links to five sections related to environmental databases providing information on:

- industry expertise;
- environmental education;
- research and development;

265

- cleaner production; and
- environmental technologies and hazardous wastes.

Each of these sections has exhaustive directories of case studies. Thus, the 'cleaner production' case studies directory details mechanisms for improving environmental and economic efficiency in industry. Each case study provides information on a particular site, including the initial problem, how it was solved using clean technology approaches, and the economic costs and benefits. The site provides useful information about industrial processes that are likely to be involved in EIAs.

Another free-access environmental database is the Right-to-Know Network (http://www.rtk.net/). It provides free access to numerous databases, text files and conferences on the environment, housing and sustainable development. It is sites such as these that could prove invaluable when preparing an environmental assessment but their relevance to the European environmental assessment market is limited because of the oversea nature of their content (Australian and North American respectively). We consider it likely that such sites will be established in the UK before long and could therefore represent a valuable source of additional information and data that is readily accessed and up to date.

Asked to search for 'environmental impact statement' and 'environmental statement' Yahoo goes to Altavista.

Altavista surfing

In contrast with Yahoo, Altavista covers almost all the information available on the Web. It is up to 100 times faster than the other search engines (and it needs to be considering the number of sites present). Its database is extensive and up to date but it is not particularly useful when a search is made for general keywords such as 'Environmental impact assessment'. The number of sites located during such a search (hundreds of thousands) makes it unrealisable to visit even a representative number of them. However, with the options of 'Advanced research' and the 'Refine' button, it is possible to narrow the search to a more manageable number of sites. Both tools are user-friendly and offer good results when you are looking for a specific topic such as the EIA of a landfill.

Two 'Advanced' search techniques were tested:

- using the binary operator NEAR which ensures that the search words you have typed are within ten words of each other in the resulting documents.
- surrounding the key phrase being searched for with double quotes, e.g. 'Environmental impact assessment'.

The several hundred thousand sites located by the simple search were reduced to thousands or even hundreds using one of the 'Advanced search' options; the two different search techniques produced similar results for the keywords/ phrases tested.

When asked to search for 'Environmental assessment' with the ranking field 'landfill', Altavista located some 5000 sites. Using the binary operator NEAR or the double quotes method reduced these to fewer than 2000 sites; many of the sites identified appeared to be duplicated when using the 'Advanced research' and 'Refine' buttons but, not surprisingly, discrepancies and differences existed. Further refining (addition of more ranking fields by identifying, for example, the type of landfill or the country of interest) reduced the number of sites identified to dozens or a few hundred. Many of these could be discounted immediately by visiting the sites, when it becames apparent they are irrelevant.

Search strategy

The following conclusions have been drawn from our experience:

- Currently most web sites are of American origin and their usefulness outside North America may be limited, but the number of Europe-based web sites is growing very rapidly.
- The rapid growth of the Internet is fuelling the exploration of information dissemination. As a consequence, it is difficult to characterise the nature, quality and the comprehensiveness of this information. The only defining characteristic of the information in the Internet is that it is transient.
- The research undertaken proved time-consuming and resulted in very few relevant documents that would feasibly be of use when undertaking an environmental assessment. When searching with the keywords, most of the web sites identified were companies offering their services, summaries of completed EIAs and of various research (universities, institutes, etc.). The last two may be of use in some circumstances.
- Most of the information found was considered to be of little value beyond casual interest; the information which might have been of use was only available for purchase.
- It pays to be very methodical during the searching process; it is very easy to branch off in pursuit of a seemingly interesting source of potential information and to be side-tracked from the purpose in hand.
- The best-quality and most available information was found to be from government web sites and from non-government organisations. These web sites included case studies and a large amount of well-organised and largely accessible data and information.
- Many web sites have links to other sites that may be more useful than the site searched, and are frequently worth a visit.
- Some databases provided reliable information that could be useful for the baseline studies both of a proposed development and of the affected environment, but they were highly specific to either that development type or the location. Information about a development could be found in

summaries of similar projects and case studies, where the details of the approach could also be reviewed, whereas measured data are located in the Environment Agency and the Department of the Environment web sites.

- Even the coarse search engines were remarkably sensitive, producing very different results for 'environmental assessment' and 'environment assessment'; when surfing, all likely variations should be included.
- The search results varied, depending on which search engines were used. Yahoo and Altavista are compared in Table 20.

Web-site addresses

Some of the more interesting and potentially useful web sites are identified below. Due to the rapidly evolving nature of the medium, no guarantee can be given of their current status, availability, usefulness or quality.

Environmental links

- *Envirolink*
  http://www.envirolink.org
  Non-profit organisation whose aim is to provide the most comprehensive, up-to-date environmental resources available, on the Internet with a link

Table 20. Comparison between search engines

Search engine	Key feature	Advantages	Limitations
Yahoo	The information has already been vetted and classified and access to the whole www does not occur. Results are general and not comprehensive; only what is available from Yahoo is obtained.	The small number of results makes review of identified sites straightforward. The taxonomy of the classification system is given, making it possible to assess the relevance.	'Advanced search' and 'Refine' options do not exist. Access to the whole www is limited.
Altavista	Comprehensively covers information available on the whole www.	The option to REFINE the search as well as the ADVANCE SEARCH.	When the keyword is too general (e.g. EIA), it is impossible to visit all the sites identified.

to every environmental site. In one of its more interesting features, is the EnviroLink Library, information is grouped under subjects within the main area Earth, Air, Fire, Water and Flora and fauna.

- *Environment on the WWW*
  http://www.ovam.be/internetrefs/english.htm
  Tries to link only to non-commercial environmental-related sites covering specific topics; updated several times a month.
- *Environmental Sites*
  htpp://www.lib.kth.se/~lg/envsite.htm
  An alphabetical subject index to environmental information on the www.
- *EnviroNet Australia*
  htpp://www.erin.gov.au/net/environet.html
  Free-access database providing information on Industry expertise, Environmental education, Research and development, Cleaner production and Environmental technologies and hazardous Wastes; each of these sections has an exhaustive directory of case studies.
- *The Right-to-Know Network*
  http://www.rtk.net
  Free access to numerous databases, text files and conferences on the environment, housing and sustainable development.
- *The Environmental Professional's Guide to the Net*
  http://www.geopac.com
  With links to environmental resources on the Internet.
- *GreenNet*
  http://www.gn.apc.org
- *EcoTradeNet*
  http://www.ecotradenet.com
  Enables identification of clean technologies, products and services and promotes environmental companies for sale of their goods and services.
- *Environmental Organisation Web Directory*
  http://www.webdirectory.com
  Free database providing comprehensive information on the Environment, in 30 sections: Agriculture, Animals, Arts, Business, Databases, Design, Disasters, Education, Employment, Energy, Forestry, General environmental interest, Government, Health, Land conservation, News and events, Parks and recreation, Pollution, Products and services, Publications, Recycling, Science, Social science, Sustainable development, Transportation, Usenet newsgroups, Vegetarianism, Water resources, Weather and Wildlife.
- *University of Virginia*
  http://ecosys.drdr.virginia.edu/Envirolists.html
  Part of the University of Virginia's Eco-web site; a powerful 'List of Lists'.
- *Visions*
  http://www.islandnet.com/~pjhughes/hope.htm

With a diverse and interesting set of links to environmental on-line resources.

- *The Global Directory for Environmental Technology*
  http://eco-web.com
  'Green Pages' site listing organisations active in environmental information, waste management and recycling.
- *The World Wide Web Virtual Library*
  http://www.ulb.ac.be/ceese/meta/sustvl.html
  The Virtual Library on Sustainable Development, maintained by the Centre for Economic and Social Studies on the Environment at l'Université Libre de Bruxelles, Belgium, includes search facilities, mailing lists, libraries and references.

National Government Organisations

United Kingdom

- *Department of the Environment, Transport and the Regions*
  http://www.dtr.gov.uk
  The DETR web site has integrated the former DoE web site information within the section identified as Environmental Protection. The Environmental Protection web page includes the following sections entitled: Air quality information; DOE/ACRE newsletters; Drinking-water information; Environmental Action Fund (EAF) 1998–99; Water Resources and Supply: Agenda for Action; Global wildlife division; Greening Government; National Environmental Health Action Plan (an overview); UK environment in facts and figures; and Indicators of Sustainable Development for the UK.

Within 'Air quality' there are details of the 'Air Quality Information Service' (information on air pollution levels across the country and multi-level maps giving estimated pollution distributions with 1 km resolution, a forecast for air quality across the UK for the next 24 hours, details of where acid rain is monitored, an explanation of the chemistry of atmospheric pollutants in the atmosphere. There is a link to the National Atmospheric Emission Inventory, which lists air pollution sources in the UK. The *'National Air Quality Information Archive'* contains data for automatic and non-automatic air quality networks within the UK (statistics back to the early 1970s) and are detailed emission inventory data. The main requirements of the local authority air pollution control (LAAPC) system established under the Environmental Protection Act 1990 are listed in *'Local Authority Pollution Controls'* with links to other sources of information, guidance notes and information on recent developments which is updated every two months.

The 'UK environment in facts and figures' covers the state of the environment in the UK, and some of the human activities which impact on it, including the topics: Global atmosphere; Air quality; Inland waters; Coastal and marine environment; Radioactivity; Noise; Waste and recycling; Land use and land cover; and Wildlife.

- *Environment Agency of England and Wales*
  http://www.environment-agency.gov.uk
  A well prepared and clearly presented site aiming to provide more information and up-to-date data on the environmental elements which the Environment Agency regulates. A key feature is the *State of the Environment* Report, a comprehensive look at the pressures on the environment and its state over the preceding 25 years.

  There is also a facility to examine information from three Environment Agency databases: Bathing water quality, River habitat survey and River gauging stations. Map images, raw data, summary statistics, text, graphics, photos and videos are offered. There are also features providing information about the Environment Agency, including its aims and objectives, functions and responsibilities.

  A section on information resources includes guidance on issues such as special waste and consignment notes, and details of other published information. There is a search engine, a comprehensive index and forms for feedback about the site.

- *The Scottish Environmental Protection Agency*
  http://www/sepa.org
- *UK Parliament*
  http://www.parliament.the-stationery-office.co.uk
- *Central Office of Information Internet Services. Government Press Releases (by Department)*
  http://www.coi.gov.uk/depts/deptslist.html

USA

- *US Environmental Protection Agency*
  http://www.epa.gov
  Comprehensive collection of resources and information covering a wide range of environmental topics. Specific pages are devoted to different groups of Internet users (concerned citizens, researchers and scientists, business and industry, etc.). The information is also organised as a set of resources covering the environmental spectrum, such as regulations, publications, research grants and environmental contracts, EPA projects and programmes, databases and software, and other sources of information.

  Within the 'Databases and software' pages, the Enviroene page attempts to provide a single repository for pollution prevention, compliance assurance, and enforcement information and data bases. It includes pollution prevention case studies, technologies, points of contact, environmental statues, executive orders, regulations, and compliance and enforcement policies and guidelines. There is also a search engine.

- *The United States Department of Energy*
  http://www.doe.gov
  Good news updates, calendar of events and background information on US fossil energy programmes. Very detailed, but well laid out and easy to use.
- *US NonProfit Gateway*
  http://www.nonprofit.gov
  A central starting point to help non-profit organisations access on-line Federal information and services.

Others

- *Environmental Resources Information Network (ERIN)*
  http://www.erin.gov.au/erin/html
  Australian Government service providing high-quality, geographically related data required for planning and decision making.
- *Wind energy information for Denmark*
  http://www.afm.dtu.dk/wind
- *International Development Research Centre*
  http://www.idrc.ca/index_e.html
  Public corporation created by the Canadian Government to help communities find solutions to environmental and other problems.

International Government Organisations

European

- *Europa*
  http://www.europa.eu.int
  Information on the EU's policies and goals with a daily news service, search facility, contact points and links to other government sites.
- *The European Environment Agency*
  http://www.eea.dk
  Main goal is to provide the European Union and member states with high-quality environmental information, to ensure that the public is properly informed about the state of the environment and to be the primary means of co-operation within its wider network, the Environmental Information and Observation Network (EIONET). Includes documents such as newsletters, press releases and brochures, full texts or abstracts from major reports, the current Annual Work Programme, monographs, articles and papers; environmental news stories and most recent EEA projects; forthcoming events; information about EEA; overviews of EEA projects; links to other web sites; databases; and search/index functions.

- *European Commission Directorate General XI (Environment)*
  http://europa.eu.int/en/comm/dgll/dglhome.html
  Policies, speeches, documents and newsletters, e.g. strategy for methane emission reduction and the draft landfill directives.
- *The European Environmental Bureau*
  http://www.greenchannel.com/eeb
- *The European Environmental Law Homepage*
  http://www.unimass.nl/~egmilieu

Non-European

- *The Organisation for Economic Co-operation and Development (OECD)*
  http://www.oecd.org
  Many statistical and factual data — activities, news, events, publications.
- *Global Legislators Organisation for a Balanced Environment (GLOBE)*
  http://globeint.org
  GLOBE, an organisation of legislators from the EU, Japan, Russia and the USA, allows parliamentary leaders to respond to global environmental challenges.
- *Global Environmental Information Locator Service (GELOS)*
  http://enrm.ceo.org
  Part of the G7 Environment and Natural Resources Management project whose main objective is to create a globally distributed virtual library.
- *United Nations Environment Programme*
  http://www.unep.org
  Six main sections: programmes (divided into categories: natural resources, production and consumption, human health and well being, globalisation trends, and global and regional servicing); products; services (including a UNEP searcher); events and activities; and a search engine.
- *UNEP International Environmental Technology Centre*
  http://www.unep.or.jp
  UNEP created an International Environmental Technology Centre to strengthen its role in the management of large cities and freshwater resources.
- *UNEP Working Group on Sustainable Product Development*
  http://unep.frw.uva.nl
  Group working in the area of sustainable products and services, offering a database and the magazine *Way Beyond*.

Non-Government Organisations

Non-profit-making NGOs

- *The World Health Organisation*
  http://www.who.int
  Excellent starting point for a wide range of environmental and health-related material.

- *Friends of the Earth*
  http://www.foe.co.uk
- Greenpeace
  http://www.greenpeace.org
- *Institute of Environmental Management (IEM)*
  http://www.iem.org.uk
  Discussion forum, IEM news, documents and information on courses which are part of IEM's signature of commitment scheme.
- *The Institution of Civil Engineers*
  http://www.ice.org.uk
- *Chartered Institute of Transport*
  http://www.citrans.org.uk
- *Institution of Highways and Transportation*
  http://www.iht.org
- *The United States Association of Water Technologies*
  http://www.awt.org
- *Air and Waste Management Association*
  http://www.awma.org
- *Construction Industry Environmental Forum*
  http://www.ciria.org.uk
  This Forum is part of the Construction Industry Research and Information Association (CIRIA), which aims to improve the performance of all concerned with construction and the environment. Offers background information on research in: Building design and materials; Construction; Management and IT; Productivity; Ground engineering; Water engineering.
- *European Energy-from-Waste Coalition*
  http://www.eewc.org
- *Asian Technology Information Program*
  http://www.atip.org
- *Japan Information Network*
  http://www.jinjapan.org/index-f.html

Others

- *Great Lakes Environmental Wire*
  http://www.cic.net/glew
- *World Forest Institute*
  http://www.vpm.com/wfi
- *Transport Planning Society*
  http://www.farabi.co.uk/tps
- *US Army Corps of Engineers*
  http://www.usace.mil
- *American Society of Naval Engineers*
  http://www.jhuapl.edu/ASNE

- *American Society of Electrical Engineers*
  http://www.webplus.net/asme
- *Environmental Routenet*
  http://www.csa.com/routenet/newaccess.html
  Subscription-based site providing a single access point for a wide array of global environmental information. The sections include: Global research (Keyword search of all RouteNet's databases); Roadmap (tabular overview of the web site); Daily news (from key sources, e.g. CNN); Data resources (links to other web sites and environment and pollution databases); Site specific information; Patents and standards; Grants and funding; Legislation and regulations; Discussion room; Market place (links to associations, consulting firms, directories of companies, and products); and Education forum (details of courses, seminars and jobs).
- *Anglian Water*
  http://www.anglianwater.co.uk
  Good environmental information as well as details of company and industry projects.
- *Contaminated Land*
  http://www.contaminatedland.co.uk
  Includes lists of contaminants with trigger levels, typical contamination from various industrial uses, and details of publications, legislation and guidance.
- *British Wind Energy Association*
  http://www.bwea.com
  Concentrates on providing as much information and as many contacts as possible on wind energy and its application. The home page has links to seven sections: New releases; Company directory; Publications and factsheets; the Annual Conference; information about the BWEA; Job vacancies; and links to other sites dealing with wind energy and other related topics. The pages are also sprinkled with links.
- *Dr Dave Buckley's Air Quality Services*
  http://www.cityscape.co.uk/users/bwø7/index.html
- *3E Database*
  http://www.nhbs.co.uk/3c/index.html
- *Netcen's air quality pages*
  http://www.aeat.co.uk/netsen/aqarchive
  A national air-quality archive containing data from the national network of monitoring sites including information on ozone, nitrogen dioxide, carbon monoxide, sulphur dioxide particulates and hydrocarbons.
- *National Society for Clean Air*
  http://www.greenchannel/com/nsca
- *Acid Rain Information Centre*
  http://www.doc.umu.ac.uk/aric/index.html
  Operated by the Manchester Metropolitan University to promote their

services, a great deal of free and useful information addressing various air-quality issues.

- *New Scientist*
  http://www.newscientist.com
  Global environmental news issues.
- *OVAM*
  http://www.ovam.be/multilang.html
  OVAM is the public waste agency of Flanders, Belgium. Well stocked with English-language material, including waste management in Belgium, contaminated land policy in Flanders, and environmental resources on the www.
- *The Recycling Council of Ontario*
  http://www.web.apc.org/rco
  Non-profit organisation which promotes reduction, reuse, recycling and composting as a means of reducing society's waste.
- *Internet Consumer Recycling Guide*
  http://www.best.com/~dillon/recycle
  Information on household material recycling. Subjects covered include common kerbside materials, guides to recycling difficult materials, and reducing unwanted mail.
- *Global Recycling Network*
  http://www.grn.com
- *RECOUP*
  http://www.tecweb.com/recoup/index.htm
  General information on how plastic bottles are recycled, details of scheme operations and details on material specification, prices and markets.
- *Environment Business Magazine*
  http://www.ifi.co.uk
  Information selected from EBM, but it is also possible to browse pages from other IFI publications. Free access. Links to the leading European environmental magazines through the European Environmental Press.
- *The Environment Council*
  http://www.greenchannel.com/tec
  Comprehensive overview of the Environment Council's programmes and activities, details of its seminars, briefings, training and events, a news digest and information on membership.
- *British Telecom*
  http://www.bt.com
- *ISO 14000 INFOCENTRE*
  http://www.iso14000.com
  Home page includes background articles, a list of acronyms, business opportunities, organisations, training, discussion groups and lists of certified companies. The articles and overviews are informative for those with little knowledge of ISO 14001. Links to other relevant sites.

- *Biffa Waste Services*
  http://www.biffa.co.uk
  Includes monitoring data from the company's sites.
- *Waste and Environment Today*
  http://aeat.co.uk/netcen/wet
  Summarised results of the journal's latest reader survey, links to NETCEN
  and other parts of AEA Technology, and 'Subscriber Club' pages.
- *Environmental Bankers' Association*
  http//environlink.org/orgs/eba/whyarewehere.htm
  Facilitates the exchange of information about the environmental aspects
  of finance.
- *Cranfield University*
  http://www.cranfield.ac.uk
  Information on environmental technology research.
- *Electric Power Research Institute*
  http://www.epri.com
- *Roslin Institute, Edinburgh*
  http://www.ri.bbsrc.ac.uk
  Research centre for molecular and quantitative genetics of farm animals
  and poultry science.

References

Andreottola, G., Cossu, R. and Serra, R. (1989) A method for the assessment of the environmental impact of a sanitary landfill. *Sanitary Landfilling: Process, Technology and Environmental Impact*, Christensen, T. H., Cossu, R. and Stegmann, R. (Eds), Academic Press, London.

Asian Development Bank (1993) *Guidelines for Incorporation of Social Dimensions in Bank Operations*, Asian Development Bank, Manila.

Barley, J. and Fables, U. (1990) A proposed framework and database for EIA auditings. *Journal of Environmental Management* 31, 163–172.

Batey, P., Madden, M. and Schofield, G. (1993) Socio economic impact assessment of large scale projects using impact-output analysis: a case study of an airport. *Regional Studies*, **27**(3), 179–192.

Beanlands, G. E. and Duinker, P. N. (1983) *An Ecological Framework for EIA in Canada*, Federal Environmental Assessment Review Office, Quebec, Canada.

Bisset, R. and Tomlinson, P. (1988) Monitoring and auditing of impacts. In *Environmental impact assessment: theory and practice*, Wather, P. (Ed.), Unwin Hyman, London.

Bourdillon, N. (1995) Limits and Standards in Environmental Impact Assessment, Working Paper No. 164, School of Planning, Oxford Brookes University.

Bourdillon, N. *et al.* (1996) Archaeological and other material and cultural assets. In *Methods of Environmental Impact Assessment*, E. & F. N. Spon, London.

Brew, D. and Lee, N. (1996) Monitoring, environmental management plans and post project analysis. *EIA Newsletter* No.12, 10–11.

Brookes, A. (1997) Quality of environmental statements — Environment Agency perspective. *Proc. Conference on Quality of Environmental Statements*, Oxford Brookes University.

278

Brownrigg, M. (1974) *A Study of Economic Impact: The University of Stirling*, Scottish Academic Press, Edinburgh.

Buckley, R. A. (1991) How accurate are environmental impact predictions? *Ambio* 20(3–4), 161–162.

Burdge, R. (1994) *A Community Guide to Social Impact Assessment*, Social Ecology Press, Middleton.

Burdge, R. (1995) *A Conceptual Approach to Social Impact Assessment: Collection of Readings*, Social Ecology Press, Middleton.

Carpenter, R. A. and Managers, J. E. (Eds) (1989) *How to Assess Environmental Impacts on Tropical Islands and Coastal Areas*. Training Manual prepared for the South Pacific Environmental Programme.

Centre for Environmental Management and Planning (1994) *Proc. Think Tank on the Effectiveness of EA*, University of Aberdeen.

Cernea, M. M. (1988) Involuntary resettlement in Development Projects: Policy Guidelines in World Bank financed projects. World Bank Technical Paper No. 80, World Bank, Washington DC.

Chapman, T. (1996) Environmental monitoring and audit. *Environmental Assessment* 4(4), 145–147.

CIRIA (1993) *Environmental Handbook for Building and Civil Engineering Projects — Construction Phase*, Special Publication 98, CIRIA, London.

CIRIA (1993b) *Environmental Handbook for Building and Civil Engineering*, 1, 2, CIRIA, London.

Clark, B. D. (1987) *Impact Assessment, Town and Country Planning Summer School. Report of Proceedings*, RTP1.

Clark, B. D. (1994) Improving public participation in EIA. *Built Environment* 20(4), 294–308.

Clark, B. D., Chapman, K., Bisset, R., Wathern, P. and Barret, M. (1981) *A Manual for the Assessment of Major Development Proposals*, HMSO, London.

Coles, T. F. and Tarling, J. P. (1991) *Environmental Assessment: Experience to Date*, Institute of Environmental Assessment, Lincoln.

Committee on the Challenges of Modern Society (1995) *Evaluation of Public Participation on EIA*, Report 207, NATO, Brussels.

Council for the Protection of Rural England (1990) *Environmental Statements — Getting Them Right*, Fact Sheet, CPRE, London.

Countryside Commission (1991) *Environmental Assessment: The Treatment of Landscape and Country Recreation Issues.*

Cox, P. R. (1993) *Badgers on Site — A Guide for Developers and Planners*, Berkshire County Council, Reading.

Daily Telegraph (1996) Great newt hunt to save 5000 homes (3 August).

Davies, M. (1990) *Screening and Scoping: Approaches to the Determination Of Need for and Scope of EIA in Environmental Impact Assessment: A Handbook*, O'Sullivan, M. (Ed.), Resource and Environmental Management Unit, University College, Cork.

Dee, N., Drobny, N. L. and Duke, K. (1973) *An Environmental Evaluation System*

279

for *Water Resource Planning*, Water Resources Research, 9, No. 3, June 1997, pp.523–553.

Department of the Environment (1988) Circular 15/88 (Welsh Office 23/88), *Environmental Assessment*, HMSO, London (in this book referred to as the 'Circular').

Department of the Environment (1990a) PPG16 Archaeology and Planning, HMSO, London.

Department of the Environment (1990b) *This Common Inheritance*, HMSO, London.

Department of the Environment (1992) *The UK Environment*, HMSO, London.

Department of the Environment (1993) *DoE Minerals Planning Guidance Note 11: The Control of Noise at Surface Mineral Workings*, HMSO, London.

Department of the Environment (1994a) PPG23 *Planning and Pollution Control*, HMSO, London.

Department of the Environment (1994b) *A Guide to the Procedures*, HMSO, London.

Department of the Environment (1994c) *Draft Guide on Preparing Environmental Statements on Planning Projects*, HMSO, London.

Department of the Environment (1994d) Circular 7/94 (Welsh Office 20/94), *Environmental Assessment: Amendment of Regulations*, HMSO, London.

Department of the Environment (1995a) *A Good Practice Guide*, HMSO, London.

Department of the Environment (1995b) *Preparation of Environmental Statements for Planning Projects that Require Environmental Assessment*, HMSO, London.

Department of the Environment (1995c) *The Environmental Effects of Dust from Surface Mineral Workings*, London, HMSO.

Department of the Environment (1996) *Validation of the UK-ADMS Dispersion Model and Assessment of its Performance Relative to R-91 and ISC Using Archived LIDAR Data*, London, HMSO.

Department of the Environment and Welsh Office (1989) *Environmental Assessment — A Guide to the Procedures*, HMSO, London (in this book referred to as the 'Blue Book').

Department of the Environment, Transport and the Regions (1997) *Mitigation Measures in Environmental Statements*, DETR, Rotherham.

Department of Transport (1983) *Manual of Environmental Appraisal*, HMSO, London.

Department of Transport (1988) *Calculation of road traffic noise*, HMSO, London.

Department of Transport (1993) *Design manual for roads and bridges volume 11: Environmental assessment*, HMSO, London.

Department of Transport (1995) *Manual of Environmental Appraisal*, HMSO, London.

ECC (1996) *Environmental Assessment — The Way Forward*.

ECC International and Rust Environmental (1995) Gaverigan china clay tip — environmental statement.

English Nature (1994) *Nature Conservation in Environmental Assessment*, English Nature, Peterborough.

English Nature (1995) *English Nature Environmental Assessment. A Guide to Best Practice*, English Nature, Peterborough.

Environment Agency (1996) *Environmental Assessment: Scoping Handbook for Projects*, HMSO, London.

Erickson, P. A. (1994) *A Practical Guide to Environmental Impact Assessment*, Academic Press, San Diego CA.

Formby, J. (1990) The politics of environmental assessment. *EIA Review* **7**(3), 207–226.

Fritz, E. S. (1980) *Strategy for Assessing Impacts of Power Plants on Fish and Shellfish Populations*, FWS/OBS–80/34, Fish and Wildlife Service, Department of the Interior, Ann Arbor, MI.

Fuller, K. and Sadler, B. (1996) The future directions of environmental assessment. *Proc. IEA/EARA Joint Annual Conference*, 24–25 Sept. 1996.

Gilpin, A. (1995) *Environmental Impact Assessment*, Cambridge University Press, Cambridge.

Glasson, J. (1992) *An Introduction to Regional Planning*, 2nd edn., UCL Press, London.

Glasson, J. (1995) Socio-economic impacts: overview and economic impacts. In *Methods of Environmental Impact Assessment*, Morris, P. and Therivel, R. (Eds), UCL Press, London.

Glasson, J. (1997) Learning from experience: changes in the quality of environmental impact statements for UK planning projects. *Proc. Conference on Changes in the Quality of EIA*, Oxford Brookes University, February 1997.

Glasson, J., Therivel, R., Weston, J., Wilson, E. and Frost, R. (1996) Changes in the quality of environmental statements for planning projects. *Environmental Assessment* **4**(3), 96–97.

Goodey, B. (1996) Landscape. In *Methods of Environmental Impact Assessment*, UCL Press, London.

Gould, R. (1996) Air quality modelling. *Croner's Environmental Management* **12** (February).

Harding, P. T. (1992) *Biological Recording of Changes in Wildlife*, HMSO, London.

HC Written Answers (1992), ENDS Report No. 208, May

Holdgate, M. W. (1979) *A Perspective of Environmental Pollution*, Cambridge University Press, Cambridge.

IEA (1993) *Guidelines for the Environmental Assessment of Road Traffic*, IEA, East Kirkby, Lincs.

IEA (1995) *Guidelines for Baseline Ecological Assessment*, Institute of Environmental Assessment, E. & F. N. Spon, London.

IEA and Landscape Institute (1995) *Guidelines for Landscape and Visual Impact Assessment*, E. & F. N. Spon, London.

IHT (1994) *Guidelines for Traffic Impact Assessment.*

Interorganisational Committee on Guidelines and Principles for SIA (1994) *Impact Assessment* **12**(2), 107–152.

Interorganisational Committee on Guidelines and Principles for SIA (1997) *Impact Assessment Review* **15**(1), 11–43.

Jolley, T. J. and Wheater, H. S. (1996) A larger-scale grid based on hydrological model of the Severn and Thames catchments. *J. CIWEM* **10** (August) 253–261.

Keith, L. H. (Ed.) (1996) *Principles of Environmental Sampling*, American Chemical Society, Washington DC.

Kletz, T. A. (1992) *Hazop and Hazon: Notes on the Identification and Assessment of Hazards*, ICE, Rugby.

Lee, N. and Lewis, M. (1991) *Environmental Assessment Guide for Passenger Transport Schemes* (for the Passenger Transport Executive Group), EIA Centre, University of Manchester.

Leistritz, L. (1995) Economic and fiscal impact assessment. *In Environmental and Fiscal Impact Assessment*, Vanclay, F. and Bronstein, D. A. (Eds), John Wiley, Chichester.

Leopold, L. B., Clarke, F. E. and Hornshaw, B. R. (1971) *A Procedure for Evaluating Environmental Impact.* Geological Survey Circular 645, Government Printing Office, Washington DC.

Mitchell, J. (1997) Mitigation in environmental assessment — furthering best practice. *Environmental Assessment* **5**(4), 28–29.

Morris, P. (1996), *Ecology — an overview. In Methods of Environmental Impact Assessment*, Morris, P. and Therivel, R. (Eds), UCL Press, London.

Munn, R. E. (1979) *Environmental Impact Assessment*, 2nd Scope Report No. 5, Wiley, Chichester.

Overseas Development Administration (1993) Social Development Handbook. A Guide to Social Issues. In *ODA Projects and Programme*, UKODA, London.

Oxford Brookes University (1997) *Environmental Impact Assessment Forum: The Implications of the Amended EIA Directive*, held at the School of Planning, 16 October 1997.

Petts, J. (1996) Environmental statement quality: experience from independent review. IEA Annual Conference, 24 September.

Petts, J. and Eduljee, G. (1994) *Environmental Impact Assessment for Waste Treatment and Disposal Facilities*, Wiley, Chichester.

Rust Environmental (1995) Environmental statement for a clinical waste incineration in Teeside, undertaken for Teeside Thermal Processing Ltd.

Rust Environmental (1996a) Environmental statement for a clinical waste incinerator for Motherwell Bridge Envirotec.

Rust Environmental (1996b) Gozo Waste Transfer Station environmental statement.

Sadler, B. (1996) *Environmental Assessment in a Changing World*, Final Report of the International Study on the Effectiveness of Environmental Assessment

Schnoor, J. (1996) *Environmental Modelling: Fate and Transport of Pollutants in Water, Soil and Air.*

Selman, P. (1992) *Environmental Planning: The Conservation and Development of Biophysical Resources*, PCP, London.

Sewell, W. R. D. and Coppock, J. T. (Eds), *Public Participation in Planning*, Wiley, Chichester.

Sheate, W. (1994) *Making an Impact. A Guide to EIA Law and Policy*, Cameron May, London

Sorenson, J. (1971) *A Framework for the Identification and Control of Resource Degradations and Conflict in the Multiple Use of the Coastal Zone*, Department of Landscape Architecture, University of California, CA, USA.

States, J. B. (1978) *A Systems Approach to Ecological Baseline Studies*, FWS/OBS–78/21, Fish and Wildlife Service, Department of the Interior, Fort Collins, CO, USA.

Treweek, J. R., Thompson, S., Veitch, N. and Japp, C. (1993) Ecological assessment of proposed road developments: a review of environmental statements. *Journal of Environmental Planning and Management* **36**, 295–307.

Tsoskounglou, E. (1997) Social Environmental Impact Assessment: A Tool for Community Sensitive and Sustainable Development, *Environment Assessment* **5**(4), 18–20.

Turner, S. (1996) Noise Impact Assessment Working Party. *Environmental Assessment* **4**(3), pp. 98–99.

University of Manchester (EIA Centre) (1995a) *Five Year Review of the Implementation of the EIA Directive*, EIA Leaflet Series No. 14.

University of Manchester (EIA Centre) (1995b) *Consultation and Public Participation Within EIA*, EIA Leaflet Series No. 10.

USAEC (1973) *Preparation of Environmental Reports for Nuclear Power Plants.* Directorate of Regulatory Standards, Regulatory Guide 4.2, UAEC Regulatory Guide Series, Washington DC.

Wates, J. (1990) *Evaluation of EIA in Ireland: A Watchdog's Perspective in EIA: A Handbook*, O'Sullivan, M. (Ed.), Resource and Environmental Management Unit, University College, Cork.

Wiesner, D. L. (1995) *The EIA Process: What it is and what it means to you: a manual for everyone concerned about the environment and decisions made about its development*, Prism Press, Dorset

Wood, C. and Jones, C. (1991) *Monitoring Environmental Assessment and Planning*, Department of the Environment Report, HMSO, London.

Wood, E. F., Sivapalon, M. and Bevan, K. (1990) Similarity and scale in catchment storm response. *Reviews of Geophysics* **28**, 1–18.

World Bank (1991) Technical Paper No. 139, *Environmental Sourcebook Volume 1*, Environment Department, World Bank, Washington DC.

World Bank (1991) *Annex C — Operational Directive 4.01 Environmental Assessment*, World Bank, Washington DC.

World Bank (1993) Public involvement in environmental assessment; requirements, opportunities and issues. *Environmental Assessment Sourcebook Volume 1*, Update 5, Environment Department, World Bank, Washington DC.

Further Reading

Books

Bakkenist, G. (1994) *Environmental Information Law, Policy and Experience*, Chapter 11, Cameron May, London.

Burnett-Hall, R. (1995) *Environmental Law*, Chapter 4, Sweet and Maxwell, London.

Commission of the European Communities (1993) Report from the Commission on the Implementation of Directive 85/337/EEC on the assessment of the effects of certain public and private projects on the environment and annex for the United Kingdom. Com 93 28 final Vol. 12, Brussels, 2 April.

Construction Industry Research and Information Association (CIRIA) (1994) *Environmental Handbook for Building and Civil Engineering Projects — Construction Phase*, Special Publication 98, CIRIA, London.

Construction Industry Research and Information Association (CIRIA) (1997) *Environmental Handbook for Building and Civil Engineering Projects — Design and Specification Phase*, Special Publication 97, CIRIA, London.

Grant, M. (Ed.) (1995) *Encyclopaedia of Planning Law and Practice*, Sweet and Maxwell, London.

Institute of Environmental Assessment (1993), *Digest of Environmental Statements*, Sweet and Maxwell, London.

Haigh, N. (1995) *Manual of Environmental Policy: the EC and Britain*, Catermill, London.

Institute of Environmental Assessment (1995) *Guidelines for Baseline Ecological Assessment*, E. & F. N. Spon, London.

Institute of Environmental Assessment and the Landscape Institute (1995) *Guidelines for Landscape and Visual Impact Assessment*, E. & F. N. Spon, London.

Sheate, W. (1994) *Making an Impact. A Guide to EIA Law and Policy*, Cameron May, London.

Sheate, W. (1995) *Making an Impact — A Guide to EA Law and Policy*, Cameron May, London.

Tomlinson, P. (1990) Buildings and health. In *Environmental Assessment and Management*, Chapter D1, Curwell, S., March, C. and Venables, R. (Eds), RIBA Publications, London.

Wood, C. (1995) *Environmental Impact Assessment: A Comparative Review*, Longman Scientific and Technical, London.

Journals

Environmental Assessment. The magazine of the Institute of Environmental Assessment.

Official Guidance

Department of the Environment (1988) Circular 15/88 (Welsh Office 23/88), *Environmental Assessment*, HMSO, London (in this book referred to as the 'Circular').

Department of the Environment (1988) Circular 24/88 (Welsh Office 48/88). *Environmental Assessment of Projects in Simplified Planning Zones and Enterprise Zones*, HMSO, London.

Department of the Environment (1994) Circular 7/94 (Welsh Office 20/94), *Environmental Assessment: Amendment of Regulations*, HMSO, London.

Department of the Environment (1995) Circular 3/95 (Welsh Office 12/95). *Permitted Development and Environmental Assessment*, HMSO, London.

Other Guidance

Crown Estate Office (1988) *Environmental Assessment of Marine Salmon Farms*, HMSO, London.

Department of the Environment (1994) Planning Research Programme. *Evaluation of Environmental Information for Planning Projects — A Good Practice Guide*, HMSO, London.

Department of the Environment (1994) Planning Research Programme. *Good Practice on the Evaluation of Environmental Projects*, Research Report, HMSO, London.

Department of the Environment (1994) Planning Research Programme. *Monitoring Environmental Assessment and Planning*, HMSO, London.

Department of the Environment (1994) PPG 23 *Planning and Pollution Control*, HMSO, London.

Department of the Environment (1995) *Preparation of Environmental Statements for Planning Projects that Require Environmental Assessment*, HMSO, London.

Department of the Environment (1996) Planning Research Programme. *Changes in the Quality of Environmental Statements for Planning Projects*, HMSO, London.

Department of the Environment and Welsh Office (1989) *Environmental Assessment — A Guide to the Procedures*, HMSO, London (in this book referred to as the 'Blue Book').

Department of the Environment and Welsh Office (1995) *Your Permitted Development Rights and Environmental Assessment*, HMSO, London.

Department of Trade and Industry (1992) *Guidelines for the Environmental Assessment of Cross Country Pipelines*, HMSO, London.

Department of Transport (1989) Departmental Standard Notice HD/8/88 *Environmental Assessment under EC Directive 85/337*, HMSO, London.

Department of Transport (1992) *Transport of Works Act 1992 : A Guide to Procedures for obtaining orders relating to Transport Systems, Inland Waterways and Works Interfering with Rights of Navigation*, HMSO, London.

Environment Agency (1996) *Environment Assessment: Scoping Handbook for Projects*, HMSO, London.

Forestry Commission (1988) *Environmental Assessment of Afforestation Projects*, HMSO, London.

Her Majesty's Inspectorate of Pollution (1995) *Planning Liaison with Local Authorities*, HMSO, London.

Index